Lecture Notes in Statistics

Edited by J. Berger, S. Fienberg, J. Gani, K. Krickeberg, and B. Singer

53

Barry C. Arnold
N. Balakrishnan

Relations, Bounds and Approximations for Order Statistics

Springer-Verlag
New York Berlin Heidelberg London Paris Tokyo

Authors

Barry C. Arnold
Department of Statistics, University of California
Riverside, CA 92521, USA

Narayanaswamy Balakrishnan
Department of Mathematics and Statistics, McMaster University
Hamilton, Ontario L8S 4K1, Canada

Mathematical Subject Classification Code (1980): 62Q05, 62G05, 62G25, 62G30, 62G35, 62G99

ISBN-13: 978-0-387-96975-6 e-ISBN-13: 978-1-4612-3644-3
DOI: 10.1007/978-1-4612-3644-3

Softcover reprint of the hardcover 1st edition 1989

2847/3140-543210

To Carole and Colleen

Preface

Bounds on moments of order statistics have been of interest since Sir Francis Galton (1902) first addressed the problem of fairly dividing first and second prize money in a competition. The present compendium of results represents our effort to sort the plethora of results into some semblance of order. We have tried to assign priority for results appropriately. We will cheerfully accept corrections. Omissions of interesting results have inevitably occurred. On this too we await (cheerful) corrections.

We are grateful to Peggy Franklin (University of California), Janet Leach, Domenica Calabria and Patsy Chan (McMaster University) who shared the responsibility of typing the manuscript. The final form of the manuscript owes much to their skill and patience.

Barry C. Arnold
Riverside, California
U. S. A.

N. Balakrishnan
Hamilton, Ontario
Canada

November, 1988

Table of Contents

List of Tables

CHAPTER 1

THE DISTRIBUTION OF ORDER STATISTICS

Let $X_1, X_2, \ldots, X_n$ denote n jointly distributed random variables. The corresponding variational series or set of order statistics is just the X's arranged in non–decreasing order. We denote the i'th smallest of the X_i's by $X_{i:n}$ ($X_{1:n}$ is the smallest, $X_{2:n}$ the second smallest, etc.). By construction

$$X_{1:n} \le X_{2:n} \le \ldots \le X_{n:n}. \tag{1.1}$$

The definition of order statistics does not require that the X_i's be independent or identically distributed. Much of the literature however focusses on the case in which the X_i's constitute a random sample from some distribution F (i.e. the i.i.d. case). We will follow that tradition but on occasion will present results appropriate for the dependent and/or non–identically distributed case.

If $X_1, \ldots, X_n$ are i.i.d. with common distribution F, it is not difficult to determine the distribution of an individual order statistic. For any i = 1,2,...,n and any $x \in \mathbb{R}$ we have

$$\begin{aligned} F_{X_{i:n}}(x) &= P(X_{i:n} \le x) \\ &= P(\text{at least i of the X's are} \le x) \\ &= \sum_{j=i}^{n} \binom{n}{j} [F(x)]^j [1 - F(x)]^{n-j}. \end{aligned} \tag{1.2}$$

An alternative representation of (1.2) is possible. Introduce the notation $U_1, U_2, \ldots, U_n$ to represent a sample from a uniform (0,1) distribution with order statistics $U_{1:n}, U_{2:n}, \ldots, U_{n:n}$. For any distribution function F we define its corresponding inverse distribution function or quantile function F^{-1} by

$$F^{-1}(y) = \sup[x\colon F(x) \le y]. \tag{1.3}$$

It is readily verified that if $X_1, \ldots, X_n$ are i.i.d. with common distribution function F, then

$$F^{-1}(U_i) \stackrel{d}{=} X_i \tag{1.4}$$

and

$$F^{-1}(U_{i:n}) \stackrel{d}{=} X_{i:n}. \tag{1.5}$$

Now the distribution of $U_{i:n}$ is

$$F_{U_{i:n}}(u) = \sum_{j=i}^{n} \binom{n}{j} u^j (1-u)^{n-j}, \qquad 0 < u < 1. \tag{1.6}$$

This is clearly absolutely continuous with corresponding density

$$f_{U_{i:n}}(u) = \frac{n!}{(i-1)!\,(n-i)!} u^{i-1} (1-u)^{n-i}, \qquad 0 < u < 1. \tag{1.7}$$

The distribution function can then be written

$$F_{U_{i:n}}(u) = \int_0^u \frac{n!}{(i-1)!\,(n-i)!} t^{i-1} (1-t)^{n-i}\, dt. \tag{1.8}$$

The relation (1.5) or comparison of (1.2), (1.6) and (1.8) yields the general expression

$$F_{X_{i:n}}(x) = \int_0^{F(x)} \frac{n!}{(i-1)!(n-i)!} t^{i-1}(1-t)^{n-i}\, dt. \tag{1.9}$$

Expression (1.9) is valid for any common parent distribution F(x) for the X_i's. Only if F(x) is everywhere differentiable can we use the chain rule and obtain a density for $X_{i:n}$. In such an absolutely continuous case we have

$$f_{X_{i:n}}(x) = \frac{n!}{(i-1)!(n-i)!} [F(x)]^{i-1}[1-F(x)]^{n-i}f(x). \tag{1.10}$$

See Exercise 1 for an alternative derivation of (1.10). It is possible to write down the joint density of two or more order statistics corresponding to a sample from an absolutely continuous distribution F with density f. For example, if $1 \le i < j \le n$ we have

$$f_{X_{i:n},X_{j:n}}(x_{i:n},x_{j:n}) = \frac{n!}{(i-1)!(j-i-1)!(n-j)!} [F(x_{i:n})]^{i-1}[F(x_{j:n}) - F(x_{i:n})]^{j-i-1}$$

$$[1 - F(x_{j:n})]^{n-j} f(x_{i:n})f(x_{j:n}), \quad -\infty < x_{i:n} < x_{j:n} < \infty. \tag{1.11}$$

Alternatively it is evident that the joint density of all n order statistics is

$$f_{X_{1:n},\ldots,X_{n:n}}(x_{1:n},\ldots,x_{n:n}) = n! \prod_{i=1}^{n} f(x_{i:n}), \quad -\infty < x_{1:n} < x_{2:n} < \ldots < x_{n:n} < \infty. \tag{1.12}$$

Judicious integration yields (1.10) and (1.11). Evaluation of expectations of order statistics or of functions of order statistics is frequently most expeditiously performed using the representation (1.5). The mean of $X_{i:n}$ is thus

$$\mu_{i:n} = E(X_{i:n})$$

$$= \int_0^1 F^{-1}(u)f_{U_{i:n}}(u)\, du. \tag{1.13}$$

Analytic expressions for $\mu_{i:n}$ are rarely obtainable. The exceptional cases are discussed in Exercise 3.

The distribution of linear combinations of order statistics are often of interest since they frequently provide robust estimates of important parametric functions. Unfortunately it is rarely possible to determine other than asymptotic distributions. A case of particular prominence is the sample range defined by

$$R_n = X_{n:n} - X_{1:n}. \tag{1.14}$$

In the absolutely continuous case we can obtain an expression for the density of R_n by making a suitable transformation of the joint density of $(X_{1:n},X_{n:n})$. In this manner we obtain

$$f_{R_n}(r) = n(n-1) \int_{-\infty}^{\infty} [F(x+r) - F(x)]^{n-2} f(x)f(x+r)dx. \tag{1.15}$$

The integration described in (1.15) is almost never tractable. The lone exception involves the important special case where the X_i's are uniform (a,b). In that case we can verify that $R_n/(b-a)$ has a Beta (n–1,2) distribution (see also Exercise 5).

In the case where F is continuous (so ties are impossible) a useful Markov chain representation is possible for the order statistics. Elementary computations verify that (for $i < j$) the conditional distribution of $X_{j:n}$ given $X_{i:n} = x$ is the same as the unconditional distribution of the random variable $Y_{j-i:n-i}$ where the distribution of the Y_i's is that of the X_i's truncated on the left at x. It is also readily verified that the sequence $X_{1:n},X_{2:n},\ldots,X_{n:n}$ does indeed constitute a Markov chain. These observations are particularly useful in cases in which left–truncated versions of the distribution of X are tractable (e.g. the exponential and classical Pareto cases). The exponential case is particularly friendly. Since minima of independent exponential random variables are again exponentially distributed and since the exponential distribution has the renowned lack of memory property one may verify that the set of exponential spacings $Y_{1:n},\ldots,Y_{n:n}$ where

$$Y_{i:n} = X_{i:n} - X_{i-1:n} \tag{1.16}$$

(in which $X_{0:n} \equiv 0$) constitute a set of n independent random variables with

$$Y_{i:n} \sim \exp\left[(n - i + 1)\lambda\right]. \tag{1.17}$$

This result which also can be obtained via a simple Jacobian argument beginning with the joint density of $X_{1:n},\ldots,X_{n:n}$ is usually attributed to P.V. Sukhatme (1937).

The remainder of this monograph will focus on the wide range of results extant dealing with moments of order statistics (plus a few related excursions). The basic reference for order statistics remains H.A. David's (1981) encyclopedic treatment. The reader of the present monograph will undoubtedly wish to have that volume close at hand. With this in mind, we have endeavored to modify only slightly the notation used in that book. The only major changes involve the use of F(x) rather than P(x) for the common distribution of the sample X's and the use of $F^{-1}(u)$ rather than x(u) for the inverse distribution or quantile function. The references at the end of the present work are just that. They represent a listing of articles and books referred to in the book. H.A. David's reference list can be regarded as a convenient almost comprehensive related bibliography.

Exercises

1. In the case where the common distribution, F, of the X_i's is absolutely continuous with density f we may obtain the density of $X_{i:n}$ by a limiting argument. Begin with the observation that

$$P(x \le X_{i:n} \le x+\delta) = P\Big[i-1 \text{ X's are less than x and exactly one X is in the interval } (x, x+\delta)\Big] + o(\delta).$$

Use this to verify (1.10).

2. Verify (1.11) using an argument analogous to that used in Exercise 1.

3. Suppose $X_1,\ldots,X_n$ are i.i.d.. In each of the following cases verify the given expression for $\mu_{i:n} = E(X_{i:n})$

(i) $X_1 \sim$ uniform (a,b), $\mu_{i:n} = a + (b-a)[i/(n+1)]$.

(ii) $X_1 \sim \exp(\lambda)$, $\mu_{i:n} = \lambda^{-1}\Big[\sum_{j=1}^{i} \frac{1}{n+1-j}\Big]$.

(iii) $P(X_1 \le x) = x^{\delta}$, $0 < x < 1$ where $\delta > 0$ (a power function distribution),

$\mu_{i:n} = B(i+\delta^{-1}, n-i+1)/B(i, n-i+1)$.

(iv) $P(X_1 > x) = (x/\sigma)^{-\alpha}$, $x > \sigma$ where $\alpha > 0$, $\sigma > 0$ (a classical Pareto distribution),

$\mu_{i:n} = \sigma\, B(i-1, n-i+1-\alpha^{-1})/B(i, n-i+1)$, provided $\alpha^{-1} < (n-i+1)$.

(v) $X_1 \sim$ Bernoulli (p) (i.e. $P(X_1 = 0) = 1-p$, $P(X_1 = 1) = p$),

$$\mu_{i:n} = \sum_{j=0}^{i} \binom{n}{j} p^j (1-p)^{n-j}.$$

4. (A recurrence formula for densities of order statistics.) Denote the density of $X_{i:n}$ by $f_{i:n}(x)$. Verify that

$$(n-i)f_{i:n}(x) + i\, f_{i+1:n}(x) = n\, f_{i:n-1}(x).$$

This result could be used to derive a recurrence relation for $\mu_{i:n}$ to be discussed in Chapter 2.

5. If $X_1,X_2,\ldots,X_n$ are i.i.d. uniform (a,b) and $1 \le i < j \le n$, determine the density of the random variable $R_{i,j:n} = X_{j:n} - X_{i:n}$.

6. (Khatri (1962)). Suppose $X_1,\ldots,X_n$ are i.i.d. non–negative integer valued random variables. Define $p_n = P(X \le n)$, $n = 0,1,2,\ldots$. Verify that for $0 \le i < j \le n$ and $k \le \ell$,

$$P(X_{i:n} = k, X_{j:n} = \ell) = \frac{n!}{(i-1)!(j-i-1)!(n-j)!} \iint_{A(k,\ell)} u^{i-1}(v-u)^{j-i-1}(1-v)^{n-j}\,du\,dv$$

where the integration is over the set

$$A(k,\ell) = \{(u,v)\colon u < v,\ p_{k-1} < u < p_k,\ p_{\ell-1} < v < p_\ell\}.$$

[Hint: you may wish to treat the cases $k = \ell$ and $k < \ell$ separately.]

CHAPTER 2

RECURRENCE RELATIONS AND IDENTITIES FOR ORDER STATISTICS

2.0. Introduction

Order statistics and their moments have been of great interest from the turn of this century since Sir Francis Galton (1902) and Karl Pearson (1902) studied the distribution of the difference of two successive order statistics. The moments of order statistics did, subsequently, assume considerable importance in the statistical literature and have been numerically tabulated extensively for several distributions. For example, one can refer to Harter (1970a,b) and David (1981) for a detailed list of these tables. Meanwhile, with the primary intention of reducing the amount of direct computation of these moments, many authors including Jones (1948), Godwin (1949), Cole (1951) and Sillitto (1951, 1964) carried out independent investigations and derived several recurrence relations and identities satisfied by these moments of order statistics. Many of these relations and identities are quite useful as they express the higher order moments in terms of the lower order moments thus making the evaluation of higher order moments easy and, in addition, provide some simple checks to test the accuracy of the computation of moments of order statistics. It was only 25 years ago, however, that Govindarajulu (1963a) nicely summarized all these results and established some more recurrence relations and identities satisfied by the single and the product moments of order statistics. He then systematically applied these results in order to determine the maximum number of single and double integrals to be evaluated for the calculation of means, variances and covariances of order statistics in a sample of size n, assuming these quantities for all sample sizes less than n to be known. By a simple generalization of one of the results of Govindarajulu (1963a), Joshi (1971) determined that for distributions symmetric about zero the number of double integrals to be evaluated for even values of n is in fact zero. Recently, Joshi and Balakrishnan (1982) established similar results for any arbitrary continuous distribution and applied them to improve over the bounds of Govindarajulu. Yet another interesting application of these recurrence relations and identities among order statistics is in establishing some combinatorial identities and this has been demonstrated by Joshi (1973) and Joshi and Balakrishnan (1981a). All these results for the moments of order statistics from an arbitrary continuous distribution have been listed and analyzed by Malik et al. (1988) in their recent expository review article on this topic. Interested readers may also refer to the article by Balakrishnan et al. (1988) for similar results on the moments of order statistics from some specific continuous distributions.

In Sections 1 through 3, we discuss some relations and identities satisfied by the single moments, product moments and covariances of order statistics from an arbitrary distribution. Results of similar nature for the case when the population distribution is symmetric about zero are presented in Section 4. For the special case of the standard normal distribution, some additional recurrence relations and identities are established in Section 5 and a few interesting applications of these results are also mentioned. In Section 6 we present the relations derived by Govindarajulu (1963b) between the moments of order statistics from a symmetric distribution and the moments of order statistics from its folded distribution (folded at the center) and also discuss the cumulative rounding error committed by using these recurrence relations. Finally, in

Section 7 we show that some of these relations for moments of order statistics continue to hold when the order statistics arise from a sample of n exchangeable variates.

2.1. Relations for single moments

If $X_{i:n}$ is the i'th order statistic from a sample of size n from the distribution function F, then using equation (1.5) we may express its k'th ($k \geq 1$) moment in the form

$$\mu_{i:n}^{(k)} = \{B(i,n-i+1)\}^{-1} \int_0^1 \{F^{-1}(u)\}^k u^{i-1}(1-u)^{n-i}du. \tag{2.1}$$

In this expression B(.,.) denotes the classical beta function and F^{-1} is the inverse distribution function corresponding to F as defined in (1.3). If F is absolutely continuous an alternative expression, using (1.10), is available:

$$\mu_{i:n}^{(k)} = \{B(i,n-i+1)\}^{-1} \int_{-\infty}^{\infty} x^k \{F(x)\}^{i-1} \{1-F(x)\}^{n-i} f(x)dx. \tag{2.2}$$

The probability integral transformation u = F(x) could be used to transform (2.2) back into the form (2.1). The form (2.1) has clear computational advantages in the derivation of recurrence relations and identities in addition to being well defined for any distribution F. It will be used throughout this Chapter.

The single moments $\mu_{i:n}^{(k)}$ satisfy the following relations and identities for any distribution F.

Relation 2.1: For $n \geq 2$ and $k \geq 1$,

$$\sum_{i=1}^{n} \mu_{i:n}^{(k)} = n\,\mu_{1:1}^{(k)}. \tag{2.3}$$

Proof. The above relation follows directly if we consider the identity

$$\sum_{i=1}^{n} X_{i:n}^k = \sum_{i=1}^{n} X_i^k$$

and take the expectation on both sides.

In particular, for k = 1 and 2, the above relation yields

$$\sum_{i=1}^{n} \mu_{i:n} = n\,\mu_{1:1}$$

and

$$\sum_{i=1}^{n} \mu_{i:n}^{(2)} = n\,\mu_{1:1}^{(2)}.$$

See also Hoeffding (1953).

Relation 2.2: For $1 \leq i \leq n-1$ and $k \geq 1$,

$$i\,\mu_{i+1:n}^{(k)} + (n-i)\,\mu_{i:n}^{(k)} = n\,\mu_{i:n-1}^{(k)}. \tag{2.4}$$

Proof. From equation (2.2) we have for $1 \leq i \leq n-1$

$$n\,\mu^{(k)}_{i:n-1} = \frac{n!}{(i-1)!(n-i-1)!}\int_0^1 \{F^{-1}(u)\}^k u^{i-1} (1-u)^{n-i-1}\,du$$

$$= \frac{n!}{(i-1)!(n-i-1)!}\int_0^1 \{F^{-1}(u)\}^k u^{i-1} (1-u)^{n-i-1} (u+1-u)\,du$$

$$= \frac{n!}{(i-1)!(n-i-1)!}\left[\int_0^1 \{F^{-1}(u)\}^k u^{i} (1-u)^{n-i-1}\,du + \int_0^1 \{F^{-1}(u)\}^k u^{i-1} (1-u)^{n-i}\,du\right]$$

$$= i\,\mu^{(k)}_{i+1:n} + (n-i)\,\mu^{(k)}_{i:n}.$$

Remark 2.3: Relation 2.2 was first derived by Cole (1951) and it is important to note that it just requires the value of the k'th moment of a single order statistic in a sample of size n in order to compute the k'th moment of the remaining n–1 order statistics, assuming of course that these moments in samples of size less than n are known. This relation was proved for the case of discrete distributions by Melnick (1964); also see Abdel–Aty (1954) and Balakrishnan (1986). Equation (2.4) was dubbed the triangle rule by Arnold and Meeden (1975) who discussed several applications in the area of characterization of distributions. Extensions of (2.4) are discussed in Arnold (1977).

Relation 2.4: For even values of n, say $n = 2m$, and $k \ge 1$,

$$\frac{1}{2}\left\{\mu^{(k)}_{m+1:2m} + \mu^{(k)}_{m:2m}\right\} = \mu^{(k)}_{m:2m-1}. \tag{2.5}$$

Proof. This relation follows directly by setting $n = 2m$ and $i = m$ in equation (2.4).

Note that for $k = 1$, in particular, Relation 2.4 simply implies that the expected value of the median in a sample of even size ($n = 2m$) is equal to the expected value of the median in a sample of odd size ($n = 2m-1$).

Relation 2.5: For $m = 1,2,\ldots,n-i$ and $k \ge 1$,

$$(n-i)^{(m)}\,\mu^{(k)}_{i:n} = \sum_{r=0}^{m} (-i)^{(r)}(n)^{(m-r)}\begin{bmatrix} m \\ r\end{bmatrix}\mu^{(k)}_{i+r:n-m+r}, \tag{2.6}$$

where $(n)^{(m)}$ denotes $n(n-1)\ldots(n-m+1)$.

Proof. From equation (2.2) we have for $1 \le m \le n-i$

$$\mu^{(k)}_{i:n} = \frac{n!}{(i-1)!(n-i)!}\int_0^1 \{F^{-1}(u)\}^k u^{i-1} (1-u)^{m} (1-u)^{n-i-m}\,du$$

$$= \frac{n!}{(i-1)!(n-i)!}\sum_{r=0}^{m} (-1)^r \begin{bmatrix} m \\ r\end{bmatrix}\int_0^1 \{F^{-1}(u)\}^k u^{i+r-1} (1-u)^{n-i-m}\,du.$$

This yields the relation in (2.6) upon simplification.

For the special case when $m = n-i$, Relation 2.5 reduces to the following result.

Relation 2.6: For $1 \le i \le n-1$ and $k \ge 1$,

$$\mu^{(k)}_{i:n} = \sum_{j=i}^{n} (-1)^{j-i}\begin{bmatrix} n \\ j\end{bmatrix}\begin{bmatrix} j-1 \\ i-1\end{bmatrix}\mu^{(k)}_{j:j}. \tag{2.7}$$

Relation 2.7: For $m = 1,2,\ldots,i-1$ and $k \ge 1$,

$$(i-1)^{(m)}\,\mu^{(k)}_{i:n} = \sum_{r=n-m}^{n} (i-m-r)^{(m-n+r)}(n)^{(n-r)}\begin{bmatrix} m \\ n-r\end{bmatrix}\mu^{(k)}_{i-m:r}. \tag{2.8}$$

Proof. From equation (2.2) we have for $1 \le m \le i-1$

$$\mu_{i:n}^{(k)} = \frac{n!}{(i-1)!(n-i)!} \int_0^1 \{F^{-1}(u)\}^k u^m u^{i-1-m} (1-u)^{n-i} du$$

$$= \frac{n!}{(i-1)!(n-i)!} \sum_{r=0}^{m} (-1)^r \binom{m}{r} \int_0^1 \{F^{-1}(u)\}^k u^{i-m-1} (1-u)^{n-i+r} du$$

which, upon simplification, yields the relation in (2.8).

For the special case when $m = i-1$, Relation 2.7 reduces to the following result.

Relation 2.8: For $2 \le i \le n$ and $k \ge 1$,

$$\mu_{i:n}^{(k)} = \sum_{j=n-i+1}^{n} (-1)^{j-n+i-1} \binom{n}{j} \binom{j-1}{n-i} \mu_{1:j}^{(k)}. \tag{2.9}$$

Relations 2.6 and 2.8 are quite useful as they express the k'th moment of the i'th order statistic in a sample of size n in terms of the k'th moment of the largest and the smallest order statistics in samples of size n and less, respectively. These relations have been derived by Srikantan (1962) and Govindarajulu (1963a). Once again we could note from these two relations that we just require the value of the k'th moment of a single order statistic (either the largest or the smallest) in a sample of size n in order to compute the k'th moment of the remaining $n-1$ order statistics, given these moments in samples of size less than n. Note that this conforms with the comment made earlier in Remark 2.3 and this is to be expected after all as both Relations 2.6 and 2.8 could be obtained by repeated application of Relation 2.2. One should be careful, however, in using these two recurrence relations as increasing values of n result in large combinatorial terms and hence in an error of large magnitude. A detailed discussion of the cumulative error propagated by the use of these relations has been made by Srikantan (1962).

As mentioned in Chapter 1 few distributions admit explicit expressions for all single moments $\mu_{i:n}^{(k)}$ $(1 \le i \le n, k \ge 1)$. For some specific distributions, however, like the gamma, Weibull, extreme- value, and geometric, after noting that the expressions for $\mu_{n:n}^{(k)}$ (or $\mu_{1:n}^{(k)}$) are easy to derive, the recurrence relations in (2.4), (2.7) and (2.9) have been used by some authors to assist in computing the single moments of all the remaining order statistics; for example, see Kimball (1947), Lieblein (1955), Gupta (1960), Margolin and Winokur (1967), and Saleh et al. (1975). These computations are often checked by using Relation 2.1. This is not meaningful, however, as pointed out by Balakrishnan and Malik (1986a) since Relation 2.1 follows automatically if any one of the recurrence relations in (2.4), (2.7) and (2.9) is applied. It is rather easy to prove this point; for example, by setting $i = 1, i = 2, \ldots, i = n-1$ in (2.7) and then adding the resulting $n-1$ equations, we get

$$\sum_{i=1}^{n-1} \mu_{i:n}^{(k)} = \binom{n}{1} \mu_{1:1}^{(k)} + \sum_{r=2}^{n-1} (-1)^{r-1} \binom{n}{r} \sum_{j=0}^{r-1} (-1)^j \binom{r-1}{j} \mu_{r:r}^{(k)} + (-1)^{n-1} \sum_{r=0}^{n-2} (-1)^r \binom{n-1}{r} \mu_{n:n}^{(k)}. \tag{2.10}$$

Now making use of the combinatorial identities

$$\sum_{j=0}^{r-1} (-1)^j \binom{r-1}{j} = 0 \quad \text{and} \quad \sum_{r=0}^{n-2} (-1)^r \binom{n-1}{r} = (-1)^n$$

in equation (2.10), we obtain

$$\sum_{i=1}^{n-1} \mu_{i:n}^{(k)} = n\,\mu_{1:1}^{(k)} - \mu_{n:n}^{(k)},$$

which simply implies that Relation 2.1 will be automatically satisfied. In other words, Relation 2.1, when used as a check, will not discover errors in the starting calculations, that is, in the $\mu_{i:i}^{(k)}$ or $\mu_{1:i}^{(k)}$ $(1 \le i \le n)$. Hence, checking the numerical computations of the single moments by Relation 2.1 is not meaningful whenever any one of equations (2.4), (2.7) and (2.9) is used in the computational procedure. However, Relation 2.1 is so simple that one might still apply it to check, at least partially, whether the other single moments have been correctly derived from the $\mu_{i:i}^{(k)}$.

Let us now write $\chi_{i:n}$ for the expected value of the difference between the (i+1)'th and the i'th order statistics in a sample of size n, that is, $\chi_{i:n} = \mu_{i+1:n} - \mu_{i:n}$ for $1 \le i \le n-1$. Distributions of the differences of two successive order statistics were first discussed by Sir Francis Galton (1902) and were investigated further by Karl Pearson (1902). Sillitto (1951) established some recurrence relations satisfied by these quantities.

Relation 2.9: For i = 2,3,...,n–1,

$$n\,\chi_{i-1:n-1} - (n-i+1)\,\chi_{i-1:n} = i\,\chi_{i:n}. \tag{2.11}$$

Proof. From (2.4), we have

$$i\,\mu_{i+1:n} + (n-i)\,\mu_{i:n} = n\,\mu_{i:n-1} \tag{2.12}$$

and

$$(i-1)\,\mu_{i:n} + (n-i+1)\,\mu_{i-1:n} = n\,\mu_{i-1:n-1}. \tag{2.13}$$

Upon subtracting (2.13) from (2.12), we get

$$i\,(\mu_{i+1:n}-\mu_{i:n}) + (n-i+1)\,(\mu_{i:n}-\mu_{i-1:n}) = n\,(\mu_{i:n-1}-\mu_{i-1:n-1}),$$

which yields the relation in (2.11) .

Relation 2.10: For m = 1,2,...,i–1,

$$\chi_{i:n} = \frac{(n)^{(m)}}{(i)^{(m)}} \sum_{r=0}^{m} (-1)^r \begin{bmatrix} m \\ r \end{bmatrix} \frac{(n-i+r)^{(r)}}{(n-m+r)^{(r)}}\,\chi_{i-m:n-m+r}. \tag{2.14}$$

This relation has been derived by Sillitto (1951) and it follows directly by a repeated application of Relation 2.9. Note that this recurrence relation expresses $\chi_{i:n}$ in terms of the χ's in samples of size n and less and of order less than i. Some similar recurrence relations are available for the expected values of the i'th quasi–range, namely, $\mu_{n-i:n}-\mu_{i+1:n}$ $\left(i = 0,1,2,\ldots,\left[\frac{n-2}{2}\right]\right)$, and are due to Govindarajulu (1963a).

2.2. Relations for product moments

Using equation (1.5), the product moment of $X_{i:n}$ and $X_{j:n}$ $(1 \le i < j \le n)$ is conveniently written in the form

$$\mu_{i,j:n} = E(X_{i:n}\,X_{j:n})$$

$$= \{B(i,j-i,n-j+1)\}^{-1} \iint_{o<u<v<1} F^{-1}(u)\, F^{-1}(v)\, u^{i-1}(v-u)^{j-i-1}\, (1-v)^{n-j}\, du\, dv, \tag{2.15}$$

where

$$B(a,b,c) = \frac{\Gamma(a)\,\Gamma(b)\Gamma(c)}{\Gamma(a+b+c)} = \frac{(a-1)!\ (b-1)!\ (c-1)!}{(a+b+c-1)!}. \tag{2.16}$$

The following recurrence relations and identities are satisfied by the product moments for any distribution F.

Relation 2.11: For $n \geq 2$,

$$\sum_{i=1}^{n} \sum_{j=1}^{n} \mu_{i,j:n} = n\, \mu_{1:1}^{(2)} + n(n-1)\, \mu_{1:1}^{2}. \tag{2.17}$$

Proof. The above relation follows immediately if we consider the identity

$$\sum_{i=1}^{n} \sum_{j=1}^{n} X_{i:n}\, X_{j:n} = \sum_{i=1}^{n} \sum_{j=1}^{n} X_i\, X_j$$

and take the expectation on both sides.

Relation 2.12: For $n \geq 2$,

$$\sum_{i=1}^{n-1} \sum_{j=i+1}^{n} \mu_{i,j:n} = \binom{n}{2}\, \mu_{1:1}^{2}. \tag{2.18}$$

Proof. This relation is easily obtained from Relation 2.11 upon using the result that

$$\sum_{i=1}^{n} \mu_{i:n}^{(2)} = n\, \mu_{1:1}^{(2)}$$

given in (2.3).

Making use of the fact that $\mu_{1,2:2} = \mu_{1:1}^{2}$, one could also rewrite relation (2.18) as (Govindarajulu, 1963a)

$$\sum_{i=1}^{n-1} \sum_{j=i+1}^{n} \mu_{i,j:n} = \binom{n}{2}\, \mu_{1,2:2}. \tag{2.19}$$

Equations (2.17) – (2.19) are very simple to use and, hence, could be applied effectively to check the accuracy of the computation of the product moments.

Relation 2.13: For $2 \leq i < j \leq n$,

$$(i-1)\, \mu_{i,j:n} + (j-i)\, \mu_{i-1,j:n} + (n-j+1)\, \mu_{i-1,j-1:n} = n\, \mu_{i-1,j-1:n-1}. \tag{2.20}$$

Proof. From equation (2.15) we have for $2 \leq i < j \leq n$

$$n\, \mu_{i-1,j-1:n-1}$$

$$= \frac{n!}{(i-2)!(j-i-1)!(n-j)!} \iint_{o<u<v<1} F^{-1}(u)\, F^{-1}(v)\, u^{i-2}\, (v-u)^{j-i-1}\, (1-v)^{n-j}\, \{u+(v-u)+(1-v)\}\, du\, dv$$

$$= \frac{n!}{(i-2)!(j-i-1)!(n-j)!} \Big\{ \iint_{o<u<v<1} F^{-1}(u)\, F^{-1}(v)\, u^{i-1}\, (v-u)^{j-i-1}\, (1-v)^{n-j}\, du\, dv$$

$$+ \iint_{o<u<v<1} F^{-1}(u)\, F^{-1}(v)\, u^{i-2}\, (v-u)^{j-i}\, (1-v)^{n-j}\, du\, dv$$

$$+ \iint\limits_{o<u<v<1} F^{-1}(u)\, F^{-1}(v)\, u^{i-2}\, (v-u)^{j-i-1}\, (1-v)^{n-j+1}\, du\, dv \Big\}$$

$$= (i-1)\, \mu_{i,j:n} + (j-i)\, \mu_{i-1,j:n} + (n-j+1)\, \mu_{i-1,j-1:n}.$$

This recurrence relation has been established by Govindarajulu (1963a) and has also been pointed out for the case of normal order statistics by Teichroew (1956). It could be noted that Relation 2.13 would enable us to calculate all the product moments $\mu_{i,j:n}$ $(1 \le i < j \le n)$ by knowing n−1 suitably chosen moments, for example, the immediate upper-diagonal product moments $\mu_{i,i+1:n}$ $(1 \le i \le n-1)$. This important application of Relation 2.13 has been noted and utilized for computational purposes by several authors including Gupta (1960), Shah (1966), Saleh et al. (1975), Joshi (1982), Balakrishnan and Joshi (1984), Balakrishnan (1985), and Balakrishnan and Puthenpura (1986). However, as pointed out by Balakrishnan and Malik (1986a), it should be noted that it will not be meaningful to check the computation of the product moments by using any of the equations (2.17) – (2.19) whenever Relation 2.13 is involved in the computational procedure. This is so because the equations (2.17) – (2.19) are automatically satisfied when Relation 2.13 is applied.

Relation 2.14: For $1 \le i < j \le n$,

$$\mu_{i,j:n} + \sum_{r=0}^{i-1} \sum_{s=0}^{n-j} (-1)^{n-r-s} \binom{n}{s} \binom{n-s}{r} \mu_{n-j-s+1,n-i-s+1:n-r-s}$$

$$= \sum_{r=1}^{j-i} (-1)^{j-i-r} \binom{n}{j-r} \binom{j-r-1}{i-1} \mu_{j-r:j-r}\, \mu_{r:n-j+r}. \qquad (2.21)$$

Proof. Consider

$$\mathcal{J} = \{B(i,j-i,n-j+1)\}^{-1} \int_0^1 \int_0^1 F^{-1}(u) F^{-1}(v)\, u^{i-1} (v-u)^{j-i-1} (1-v)^{n-j}\, du\, dv. \qquad (2.22)$$

Upon expanding $(v-u)^{j-i-1}$ in powers of u and v we get

$$\mathcal{J} = \{B(i,j-i,n-j+1)\}^{-1} \sum_{s=0}^{j-i-1} (-1)^{j-i-1-s} \binom{j-i-1}{s} \int_0^1 \int_0^1 F^{-1}(u) F^{-1}(v)\, u^{j-s-2}\, v^{s} (1-v)^{n-s}\, du\, dv$$

$$= \{B(i,j-i,n-j+1)\}^{-1} \sum_{r=1}^{j-i} (-1)^{j-i-r} \binom{j-i-1}{r-1} \int_0^1 F^{-1}(u)\, u^{j-r-1}\, du \int_0^1 F^{-1}(v)\, v^{r-1}\, (1-v)^{n-j}\, dv.$$

Now on using equation (2.2) and simplifying, we get $\mathcal{J}$ as equal to the RHS of (2.21). Further, from (2.22) we also have

$$\mathcal{J} = \{B(i,j-i,n-j+1)\}^{-1} \Big[\iint\limits_{o<u<v<1} F^{-1}(u)\, F^{-1}(v)\, u^{i-1} (v-u)^{j-i-1} (1-v)^{n-j}\, du\, dv$$

$$+ (-1)^{j-i-1} \iint\limits_{o<v<u<1} F^{-1}(u)\, F^{-1}(v)\, \{1-(1-u)\}^{i-1} (u-v)^{j-i-1} (1-v)^{n-j}\, du\, dv \Big]$$

$$= \mu_{i,j:n} + \{B(i,j-i,n-j+1)\}^{-1} \sum_{r=0}^{i-1} \sum_{s=0}^{n-j} (-1)^{n-r-s} \binom{i-1}{r} \binom{n-j}{s}$$

$$\iint\limits_{o<v<u<1} F^{-1}(u) F^{-1}(v)\, v^{n-j-s}\, (u-v)^{j-i-1} (1-u)^{i-1-r}\, du\, dv, \qquad (2.23)$$

where the last equality follows by expanding $\{1-(1-u)\}^{i-1}$ and $(1-v)^{n-j}$ in powers of $(1-u)$ and v, respectively. Using now equation (2.16) and simplifying equation (2.23), we see that $\mathcal{J}$ is also equal to the LHS of (2.21). This completes the proof of the relation.

The above relation, due to Joshi and Balakrishnan (1982), is a generalization of Joshi's (1971) result. It should be noted that equation (2.21) contains only two product moments corresponding to samples of size n, viz., $\mu_{i,j:n}$ and $\mu_{n-j+1,n-i+1:n}$. In particular, setting j = i+1 in (2.21), we obtain the following relation.

Relation 2.15: For i = 1,2,...,n–1,

$$\mu_{i,i+1:n} + (-1)^n \mu_{n-i,n-i+1:n} = \sum_{r=0}^{i-1} \sum_{s=1}^{n-i-1} (-1)^{n+1-r-s} \binom{n}{s}\binom{n-s}{r} \mu_{n-i-s,n-i-s+1:n-r-s}$$

$$+ \sum_{r=1}^{i-1} (-1)^{n-r+1} \binom{n}{r} \mu_{n-i,n-i+1:n-r} + \binom{n}{i} \mu_{i:i}\, \mu_{1:n-i}. \tag{2.24}$$

Similarly, by setting j = n–i+1 in equation (2.21), we obtain the following relation.

Relation 2.16: For i = 1,2,...,[n/2],

$$\{1+(-1)^n\} \mu_{i,n-i+1:n} = \sum_{r=0}^{i-1} \sum_{s=1}^{i-1} (-1)^{n+1-r-s} \binom{n}{s}\binom{n-s}{r} \mu_{i-s,n-i-s+1:n-r-s}$$

$$+ \sum_{r=1}^{i-1} (-1)^{n-r+1} \binom{n}{r} \mu_{i,n-i+1:n-r}$$

$$+ \sum_{r=1}^{n-2i+1} (-1)^{n-r+1} \binom{n}{i+r-1}\binom{n-i-r}{i-1} \mu_{n-i-r+1:n-i-r+1}\, \mu_{r:i+r-1}. \tag{2.25}$$

Relation 2.15 shows that when n is odd, we need to calculate only (n–1)/2 product moments $\mu_{i,i+1:n}$ $(1 \le i \le \frac{n-1}{2})$. Similarly, Relation 2.16 shows that when n is even, the product moments $\mu_{i,n-i+1:n}$ $(1 \le i \le [\frac{n}{2}])$ could all be obtained from the moments in samples of size (n–1) and less. In particular, for i = 1 and even values of n, say n = 2m, Relation 2.16 yields

$$2\, \mu_{1,2m:2m} = \sum_{r=1}^{2m-1} (-1)^{r-1} \binom{2m}{r} \mu_{r:r}\, \mu_{2m-r:2m-r}, \tag{2.26}$$

a relation that has been derived by Govindarajulu (1963a) and generalized by Joshi (1971); see also Ruben (1956a) and Balakrishnan (1982). In addition, by setting n = 2m and i = m in equation (2.25), we obtain the following recurrence relation.

Relation 2.17: For m ≥ 1,

$$2\, \mu_{m,m+1:2m} = \sum_{r=0}^{m-1} \sum_{s=1}^{m-1} (-1)^{r+s-1} \binom{2m}{s}\binom{2m-s}{r} \mu_{m-s,m-s+1:2m-r-s}$$

$$+ \sum_{r=1}^{m-1} (-1)^{r-1} \binom{2m}{r} \mu_{m,m+1:2m-r} + \binom{2m}{m} \mu_{m:m}\, \mu_{1:m}. \tag{2.27}$$

2.3. Relations for covariances

Govindarajulu (1963a) obtained upper bounds for the number of single and double integrals to be evaluated for calculating all the single and product moments in a sample of size n provided these are available in samples of sizes n–1 and less. Relations 2.15, 2.16 and 2.17 have been made use of by Joshi and Balakrishnan (1982) in improving these bounds and their result is summarized in the following theorem.

Theorem 2.18: In order to find the first and the second single moments and product moments of order statistics in a sample of size n drawn from an arbitrary distribution, given these moments in samples of sizes n–1 and less, one has to evaluate at most two single integrals and (n–2)/2 double integrals if n is even; and two single integrals and (n–1)/2 double integrals if n is odd.

Proof. In view of Relation 2.6, it is sufficient to evaluate just two single integrals $\mu_{n:n}$ and $\mu_{n:n}^{(2)}$ for calculating all the first and the second order single moments, viz., $\mu_{i:n}$ and $\mu_{i:n}^{(2)}$ $(1 \le i \le n)$. Further, Relation 2.13 enables us to calculate all the product moments $\mu_{i,j:n}$ $(1 \le i < j \le n)$ by knowing (n–1) suitably chosen moments, for example $\mu_{i,i+1:n}$ $(1 \le i \le n-1)$. When n is odd, we need to calculate only (n–1)/2 product moments $\mu_{i,i+1:n}$ $\left[1 \le i \le \frac{n-1}{2}\right]$ as the remaining moments $\mu_{i,i+1:n}$ $\left[\frac{n+1}{2} \le i \le n-1\right]$ could be obtained by using Relation 2.15. Similarly, when n is even, say n = 2m, we need to calculate only (n–2)/2 = m–1 product moments $\mu_{i,i+1:2m}$ $(1 \le i \le m-1)$ as $\mu_{m,m+1:2m}$ could be obtained from Relation 2.17, while Relation 2.15 gives the remaining moments $\mu_{i,i+1:2m}$ $(m+1 \le i \le 2m-1)$. Hence the theorem.

Relation 2.19: For $1 \le i \le n-1$ and $1 \le k \le n-i$,

$$\sum_{s=i+1}^{n-k+1} \begin{bmatrix} n-s \\ k-1 \end{bmatrix} \mu_{i,s:n} + \sum_{r=1}^{i} \sum_{s=i+1}^{i+k} \begin{bmatrix} s-r-1 \\ s-i-1 \end{bmatrix} \begin{bmatrix} n-s \\ n-k-i \end{bmatrix} \mu_{r,s:n} = \begin{bmatrix} n \\ k \end{bmatrix} \mu_{1:k}\, \mu_{i:n-k}. \tag{2.28}$$

Proof. Consider the integral

$$\mathcal{J} = \{B(i,k,n-i-k+1)\}^{-1} \int_0^1 \int_0^1 F^{-1}(u)F^{-1}(v)\, u^{i-1}\, (1-u)^{n-k-i}\, (1-v)^{k-1}\, du\, dv, \tag{2.29}$$

which gives

$$\mathcal{J} = \{B(i,k,n-i-k+1)\}^{-1} \int_0^1 F^{-1}(u)\, u^{i-1}\, (1-u)^{n-k-i}\, du \int_0^1 F^{-1}(v)\, (1-v)^{k-1}\, dv$$

$$= \begin{bmatrix} n \\ k \end{bmatrix} \mu_{1:k}\, \mu_{i:n-k}$$

on using equation (2.2). From (2.29) we next write

$$\mathcal{J} = J_1 + J_2,$$

where J_1 is the integral over the set $0 < u < v < 1$ and J_2 is the integral over the set $0 < v < u < 1$. First, let us consider J_1. Writing 1–u as (1–v) + (v–u) and expanding $(1-u)^{n-k-i}$ binomially in powers of (1–v) and (v–u), we have

$$J_1 = \{B(i,k,n-i-k+1)\}^{-1} \sum_{s=0}^{n-k-i} \begin{bmatrix} n-k-i \\ s \end{bmatrix} \iint_{0<u<v<1} F^{-1}(u)\, F^{-1}(v)\, u^{i-1}\, (v-u)^{s}\, (1-v)^{n-i-s-1}\, du\, dv$$

$$= \{B(i,k,n-i-k+1)\}^{-1} \sum_{s=0}^{n-k-i} \begin{bmatrix} n-k-i \\ s \end{bmatrix} B(i,s+1,n-i-s)\, \mu_{i,i+s+1:n}$$

on using (2.15). Simplifying this expression, J_1 reduces to the first term on the LHS of (2.28). Similarly, writing u and 1−v appearing in J_2 as v + (u−v) and (1−u) + (u−v), respectively, and expanding the terms u^{i-1} and $(1-v)^{k-1}$ binomially, we find J_2 is equal to the second term on the LHS of (2.28). This completes the proof of the relation.

By setting i = 1 in Relation 2.19, in particular, we obtain the following relation.

Relation 2.20: For $1 \le k \le n-1$,

$$\sum_{j=2}^{n-k+1} \binom{n-j}{k-1} \mu_{1,j:n} + \sum_{j=2}^{k+1} \binom{n-j}{n-k-1} \mu_{1,j:n} = \binom{n}{k} \mu_{1:k}\, \mu_{1:n-k}. \tag{2.30}$$

It could be easily noted that equation (2.30) for k is same as the relation for n−k, and hence there are only [n/2] distinct relations. The important point about the above relation is that it involves the product moments $\mu_{1,j:n}$ $(2 \le j \le n)$ and first order single moments only. Therefore, for even values of n, there are n/2 distinct relations in (n−1) product moments, and a knowledge of (n−2)/2 of these, for example, $\mu_{1,2:n}$, $\mu_{1,3:n}, \ldots, \mu_{1,\frac{n}{2}:n}$, would allow us to calculate all of these moments provided the first order single moments in samples of size n−1 and less are known. Similarly, for odd values of n, we need to know just (n−1)/2 product moments, for example, $\mu_{1,2:n}$, $\mu_{1,3:n}, \ldots, \mu_{1,\frac{n+1}{2}:n}$. Note that these bounds are exactly the same as the bounds given in Theorem 2.18 for the double integrals to be evaluated for the calculation of all the product moments. This is not surprising since the product moments $\mu_{1,j:n}$ $(2 \le j \le n)$, along with Relation 2.13, are also sufficient for the evaluation of all the product moments.

Furthermore, by setting k = 1 in Relation 2.19, we immediately obtain the following relation.

Relation 2.21: For $1 \le i \le n-1$,

$$\sum_{j=i+1}^{n} \mu_{i,j:n} + \sum_{r=1}^{i} \mu_{r,i+1:n} = n\, \mu_{1:1}\, \mu_{i:n-1}. \tag{2.31}$$

The above relation has been made use of by Joshi and Balakrishnan (1982) in establishing some similar relations for the covariances of order statistics and these are presented in the following.

Relation 2.22 : For $1 \le i \le n-1$,

$$\sum_{j=i+1}^{n} \sigma_{i,j:n} + \sum_{r=1}^{i} \sigma_{r,i+1:n} = \Big(i\, \mu_{1:1} - \sum_{r=1}^{i} \mu_{r:n}\Big)(\mu_{i+1:n} - \mu_{i:n}). \tag{2.32}$$

Proof: Since $\mu_{i,j:n} = \sigma_{i,j:n} + \mu_{i:n}\, \mu_{j:n}$, equation (2.31) immediately gives

$$\sum_{j=i+1}^{n} \sigma_{i,j:n} + \sum_{r=1}^{i} \sigma_{r,i+1:n}$$

$$= n\, \mu_{1:1}\, \mu_{i:n-1} - \mu_{i:n} \sum_{j=i+1}^{n} \mu_{j:n} - \mu_{i+1:n} \sum_{r=1}^{i} \mu_{r:n}. \tag{2.33}$$

Using the identity

$$\sum_{j=i+1}^{n} \mu_{j:n} = n\, \mu_{1:1} - \sum_{r=1}^{i} \mu_{r:n}$$

obtained from Relation 2.1, the RHS of equation (2.33) becomes

$$n\, \mu_{1:1}\, (\mu_{i:n-1} - \mu_{i:n}) - (\mu_{i+1:n} - \mu_{i:n}) \sum_{r=1}^{i} \mu_{r:n}.$$

Now upon using the result

$$n\,(\mu_{i:n-1}-\mu_{i:n}) = i\,(\mu_{i+1:n}-\mu_{i:n})$$

obtained from Relation 2.2, we derive the relation given in equation (2.32).

Relations 2.21 and 2.22 provide extremely simple and useful results for checking the calculations of product moments and covariances of order statistics in a random sample of size n. In particular, setting i = 1 and i = n–1 in Relation 2.22, we get the identities

$$2\,\sigma_{1,2:n} + \sum_{j=3}^{n} \sigma_{1,j:n} = (\mu_{1:1}-\mu_{1:n})(\mu_{2:n}-\mu_{1:n}) \tag{2.34}$$

and

$$\sum_{r=1}^{n-2} \sigma_{r,n:n} + 2\,\sigma_{n-1,n:n} = (\mu_{n:n}-\mu_{1:1})(\mu_{n:n}-\mu_{n-1:n}). \tag{2.35}$$

Both equations (2.34) and (2.35) contain the covariances appearing in only one row and one column of the variance–covariance matrix $((\sigma_{i,j:n}))$. In general, Relation 2.22 involves the elements of the i'th row and (i+1)'th column of $((\sigma_{i,j:n}))$ with the coefficient of $\sigma_{i,i+1:n}$ equal to two and the coefficients of all the remaining elements, viz. $\sigma_{i,j:n}$ $(i + 2 \le j \le n)$ and $\sigma_{i,j+1:n}$ $(1 \le i \le j-1)$, equal to one. In view of the comments made by Balakrishnan and Malik (1986a), it would also be better to check the accuracy of the computations of the covariances by using Relation 2.22 in addition to the identity

$$\sum_{i=1}^{n}\sum_{j=1}^{n} \sigma_{i,j:n} = n\,\sigma_{1,1:1} = n\left\{\mu_{1:1}^{(2)} - \mu_{1:1}^{2}\right\} \tag{2.36}$$

obtained from Relation 2.1.

2.4. Results for symmetric populations

For distributions F which are symmetric about zero, it follows from (1.5) that

$$X_{i:n} \stackrel{d}{=} (-X)_{n-i+1:n}, \qquad 1 \le i \le n,$$

$$(X_{i:n}, X_{j:n}) \stackrel{d}{=} ((-X)_{n-j+1:n}, (-X)_{n-i+1:n}), \qquad 1 \le i < j \le n.$$

Hence, for population distributions symmetric about zero, we immediately have the results

$$\mu_{n-i+1:n}^{(k)} = (-1)^k\,\mu_{i:n}^{(k)}, \qquad 1 \le i \le n,\ k \ge 1, \tag{2.37}$$

$$\mu_{n-j+1,n-i+1:n} = \mu_{i,j:n}, \qquad 1 \le i < j \le n, \tag{2.38}$$

and

$$\begin{aligned}\sigma_{n-j+1,n-i+1:n} &= \mu_{n-j+1,n-i+1:n} - \mu_{n-j+1:n}\,\mu_{n-i+1:n}\\ &= \mu_{i,j:n} - \mu_{j:n}\,\mu_{i:n}\\ &= \sigma_{i,j:n}, \qquad 1 \le i < j \le n.\end{aligned} \tag{2.39}$$

These results could be used to simplify many of the relations established in Sections 1 through 3. Finally, these recurrence relations would help us reduce the bounds given in Theorem 2.18 for the number of single and double integrals to be evaluated in order to compute the first and the second single moments and product moments in a sample of size n.

Relation 2.23: For even values of n, say n = 2m,

$$\mu_{m:2m-1}^{(k)} = 0 \quad \text{for odd } k \tag{2.40}$$

$$= \mu_{m:2m}^{(k)} \quad \text{for even } k. \tag{2.41}$$

Proof. Equation (2.40) follows directly by setting n = 2m–1 and i = m in (2.37). Equation (2.41) is obtained from Relation 2.4 simply by noting that $\mu_{m+1:2m}^{(k)} = \mu_{m:2m}^{(k)}$ for even values of k.

Relation 2.24: For $1 \le i < j \le n$,

$$(1+(-1)^n)\,\mu_{i,j:n} = \sum_{r=1}^{j-i} (-1)^{j-i-r-1} \binom{n}{j-r}\binom{j-r-1}{i-1} \mu_{1:j-r}\,\mu_{r:n-j+r}$$

$$+ \sum_{r=1}^{i-1} (-1)^{n-r-1} \binom{n}{r} \mu_{n-j+1,n-i+1:n-r}$$

$$+ \sum_{r=0}^{i-1} \sum_{s=1}^{n-j} (-1)^{n-r-s-1} \binom{n}{s}\binom{n-s}{r} \mu_{n-j-s+1,n-i-s+1:n-r-s}. \tag{2.42}$$

Proof. This relation follows at once from Relation 2.14 upon using (2.37) and (2.38).

The important point about the above relation is that the RHS of (2.42) involves only the first single moments and the product moments of order statistics in samples of sizes (n–1) and less. Therefore, for even values of n, there is no need to evaluate any double integral. In this case, an exact expression for the product moment $\mu_{i,j:n}$ is obtained from equation (2.42) as

$$2\,\mu_{i,j:n} = 2\,\mu_{n-j+1,n-i+1:n}$$

$$= \sum_{r=1}^{j-i} (-1)^{j-i-r-1} \binom{n}{j-r}\binom{j-r-1}{i-1} \mu_{1:j-r}\,\mu_{r:n-j+r} + \sum_{r=1}^{i-1} (-1)^{r-1} \binom{n}{r} \mu_{n-j+1,n-i+1:n-r}$$

$$+ \sum_{r=0}^{i-1} \sum_{s=1}^{n-j} (-1)^{r+s-1} \binom{n}{s}\binom{n-s}{r} \mu_{n-j-s+1,n-i-s+1:n-r-s}. \tag{2.43}$$

This result, due to Joshi and Balakrishnan (1982), is a simple generalization of the results of Joshi (1971) and Govindarajulu (1963a). In particular, by setting i = 1 and j = 2 in equation (2.43) and using the facts that $\mu_{1:1} = 0$ and $\mu_{1,2:2} = \mu_{1:1}^2 = 0$, we obtain for even values of n

$$2\,\mu_{1,2:n} = \sum_{s=1}^{n-2} (-1)^{s-1} \binom{n}{s} \mu_{1,2:n-s}. \tag{2.44}$$

The following theorem, an analogue of Theorem 2.18 for the case when the population distribution is symmetric about zero, makes use of equations (2.40) – (2.43) and determines upper bounds for the number of single and double integrals to be evaluated for calculating all the first two single moments and product moments in a sample of size n.

Theorem 2.25: In order to find the first and the second single moments and product moments of order statistics in a sample of size n drawn from an arbitrary distribution symmetric about zero, given these moments for all sample sizes less than n, one has to evaluate at most one single integral if n is even; and one single integral and (n–1)/2 double integrals if n is odd.

Proof. To compute the first single moments $\mu_{i:n}$ $(1 \le i \le n)$, in view of Relation 2.2 one has to evaluate at most one single integral if n is even and 0 integrals if n is odd since in this case we have $\mu_{\frac{n+1}{2}:n} = 0$ from Relation 2.23. Similarly, to compute the second single moments $\mu_{i:n}^{(2)}$ $(1 \le i \le n)$, once again in view of Relation 2.2 one has to evaluate at most one single integral if n is odd and 0 integrals if n is even since in this case we have the result $\mu_{\frac{n}{2}:n}^{(2)} = \mu_{\frac{n}{2}:n-1}^{(2)}$ from Relation 2.23. Finally, to compute the product moments $\mu_{i,j:n}$ $(1 \le i < j \le n)$, one has to evaluate at most $(n-1)/2$ double integrals if n is odd (see Theorem 2.18); however, one doesn't have to evaluate any double integral if n is even since in this case one could compute all the product moments directly from equation (2.43). This completes the proof of the theorem.

2.5. Results for normal population

In addition to the relations presented in Sections 1 through 4, order statistics from some specific continuous distributions like the normal, logistic and exponential possess some more important recurrence relations and identities for their moments. Recently, Balakrishnan et al. (1988) have listed and analyzed these results for various continuous distributions in their detailed review article. In this section, we restrict our attention to the order statistics from the normal population and derive some relations satisfied by their moments and also illustrate some interesting applications of these results.

Relation 2.26: If g(x) is any differentiable function such that differentiation of g(x) with respect to its argument and expectation of g(X) with respect to an absolutely continuous distribution are interchangeable, then for $1 \le i \le n$

$$E\{g'(X_{i:n})\} = -\sum_{j=1}^{n} E\{g(X_{i:n})\, f'(X_{j:n})/f(X_{j:n})\}, \tag{2.45}$$

where f(x) denotes the population density function.

Proof. The above relation has been established by Govindarajulu (1963a) and the method of proof presented here is originally due to Seal (1956). We have for all real t

$$\begin{aligned} E\{g(X_{i:n}+t)\} &= n! \iint\limits_{-\infty<x_1<\dots<x_n<\infty} \cdots \int g(x_i+t) \prod_{j=1}^{n} f(x_j)\, dx_j \\ &= n! \iint\limits_{-\infty<x_1<\dots<x_n<\infty} \cdots \int g(x_i) \prod_{j=1}^{n} f(x_j-t)\, dx_j. \end{aligned} \tag{2.46}$$

Differentiating both sides of equation (2.46) with respect to t and then setting t = 0, we obtain

$$E\{g'(X_{i:n})\} = n! \iint\limits_{-\infty<x_1<\dots<x_n<\infty} \cdots \int g(x_i) \left[-\sum_{j=1}^{n} \{f'(x_j)/f(x_j)\}\right] \prod_{k=1}^{n} f(x_k)\, dx_k,$$

which yields the relation in (2.45).

In particular, taking g(x) = x in Relation 2.26 we get for $1 \le i \le n$

$$\sum_{j=1}^{n} E\left\{X_{i:n}\, f'(X_{j:n})/f(X_{j:n})\right\} = -1. \tag{2.47}$$

Relation 2.27: For a standard normal population,

$$\sum_{j=1}^{n} \mu_{i,j:n} = 1, \quad 1 \le i \le n, \tag{2.48}$$

and

$$\sum_{j=1}^{n} \sigma_{i,j:n} = 1, \quad 1 \le i \le n. \tag{2.49}$$

Proof. Equation (2.48) follows directly from (2.47) upon noting that $f'(x) = -x\, f(x)$ for the standard normal distribution. Equation (2.49) is obtained from (2.48) upon using the result that $\sum_{j=1}^{n} \mu_{j:n} = n\mu_{1:1} = 0$ (see Relation 2.1).

Relation 2.27 also follows directly from the well known independence property of $\bar{X}$ and $X_{i:n}-\bar{X}$ (i = 1,2,...,n) as proved by McKay (1935).

Relation 2.28: For a standard normal population, we have for $1 \le i \le n$

$$\mu_{i:n}^{(2)} = 1 + n\binom{n-1}{i-1} \sum_{j=0}^{n-i} (-1)^j \binom{n-i}{j} \frac{1}{i+j}\, \mu_{1,2:i+j}. \tag{2.50}$$

Proof. From equation (2.2), we have for $1 \le i \le n$

$$B(i,n-i+1)\, \mu_{i:n}^{(2)} = \int_{-\infty}^{\infty} x^2\, \{F(x)\}^{i-1}\, \{1-F(x)\}^{n-i}\, f(x)\, dx. \tag{2.51}$$

Writing $x\, f(x)$ as $-f'(x)$ in equation (2.51) and then integrating by parts once, we obtain

$$\begin{aligned} B(i,n-i+1)\, \mu_{i:n}^{(2)} &= B(i,n-i+1) + (i-1) \sum_{j=0}^{n-i} (-1)^j \binom{n-i}{j} \int_{-\infty}^{\infty} x\, \{F(x)\}^{i+j-2}\, f^2(x)\, dx \\ &\quad + (n-i) \sum_{j=0}^{n-i-1} (-1)^{j+1} \binom{n-i-1}{j} \int_{-\infty}^{\infty} x\, \{F(x)\}^{i+j-1}\, f^2(x)\, dx. \end{aligned} \tag{2.52}$$

Since $y\, f(y) = -f'(y)$, we have $\int_x^{\infty} y\, f(y)\, dy = -\int_x^{\infty} f'(y)\, dy = f(x)$ and, therefore,

$$\begin{aligned} \int_{-\infty}^{\infty} x\, \{F(x)\}^k\, f^2(x)\, dx &= \int_{-\infty}^{\infty} x\, \{F(x)\}^k\, f(x) \left[\int_x^{\infty} y\, f(y)\, dy \right] dx \\ &= \iint_{-\infty<x<y<\infty} xy\, \{F(x)\}^k\, f(x)\, f(y)\, dy\, dx \\ &= \mu_{k+1,k+2:k+2}/(k+1)(k+2) \\ &= \mu_{1,2:k+2}/(k+1)(k+2) \end{aligned}$$

by using (2.38). Substituting this expression for the integrals in equation (2.52) and simplifying the resulting equation, we derive the relation in (2.50).

Setting i = n in Relation 2.28, in particular, we obtain for $n \ge 2$

$$\mu_{n:n}^{(2)} = \mu_{1:n}^{(2)} = 1 + \mu_{1,2:n}. \tag{2.53}$$

Remark 2.29: Apart from theoretical interest, Relations 2.27 and 2.28 could be used in several ways. First of all, they are quite useful for checking the computations of $\mu_{i,j:n}$ and $\sigma_{i,j:n}$ from the standard normal distribution as indicated by various authors who have extensively tabulated these quatities; see, for example, Teichroew (1956), Sarhan and Greenberg (1956), Harter (1961), Yamauti (1972), and Tietjen et al. (1977). Another interesting application of these relations in developing an algorithm for obtaining approximate values of the variances and covariances of normal order statistics has been displayed by Davis and Stephens (1977, 1978). They have essentially used equations (2.49) and (2.53) for improving the David and Johnson (1954) approximation of moments. A detailed discussion of this topic is presented in Chapter 4.

Proceeding exactly on same lines as in the proof of Relation 2.28 and noting that the condition $f'(x) = -x\, f(x)$ is satisfied by both the standard normal and the half normal (chi) distributions, Joshi and Balakrishnan (1981b) have derived the following result.

Relation 2.30: For standard normal and half normal populations,

$$\sum_{j=i}^{n} \mu_{i,j:n} = 1 + \sum_{j=i}^{n} \mu_{i-1,j:n}, \quad (1 \le i \le n), \tag{2.54}$$

and

$$\sum_{j=i+1}^{n} \mu_{i,j:n} = \sum_{j=i+1}^{n} \mu_{j:n}^{(2)} - (n-i), \quad (1 \le i \le n-1), \tag{2.55}$$

where $\mu_{0,j:n} = 0$ for all $j \ge 1$.

For the special case $i = n-1$, equation (2.55) simply reduces to the relation (2.53). In addition, equation (2.54) shows that the relation $\sum_{j=1}^{n} \mu_{1,j:n} = 1$ holds for half normal distribution as well.

Relation 2.31: For standard normal and half normal populations, we have for $1 \le i \le n$,

$$\sum_{j=1}^{n} \mu_{i,j:n} = 1 + n\, \mu_{1:1}\, \mu_{i-1:n-1} \tag{2.56}$$

and

$$\sum_{j=1}^{n} \sigma_{i,j:n} = 1 - (n-i+1)\, \mu_{1:1}(\mu_{i:n} - \mu_{i-1:n}), \tag{2.57}$$

where $\mu_{0:j} = 0$ for all $j \ge 1$.

Proof. Equation (2.56), for the case $i = 1$, has already been proved in (2.54). For $2 \le i \le n$, we have from Relation 2.21

$$\sum_{j=i}^{n} \mu_{i-1,j:n} + \sum_{j=1}^{i-1} \mu_{j,i:n} = n\, \mu_{1:1}\, \mu_{i-1:n-1} \tag{2.58}$$

for any arbitrary distribution. Equation (2.56) follows on adding equations (2.54) and (2.58). Equation (2.57) could now be proved easily by using (2.56).

Equations (2.30) and (2.31) are useful in finding an expression for the variance of the selection differential or reach statistic defined by $\Delta_k = \overline{X}_k - \overline{X}_n$, where $\overline{X}_n = \sum_{i=1}^{n} X_{i:n}/n$ and $\overline{X}_k = \sum_{i=n-k+1}^{n} X_{i:n}/k$. Many authors, including Schaeffer et al. (1970) and Burrows (1972,1975), have studied $\overline{X}_k$. For the standard normal population these authors have observed that $\nu_k = k\,\mathrm{Var}(\overline{X}_k)$ remains almost constant for the selected fraction k/n. Making use of the tables of moments prepared by Sarhan and Greenberg (1956), the earlier authors have tabulated ν_k for sample sizes up to twenty and all choices of k. Burrows (1972,1975), on the other hand, has provided approximations to $E(\overline{X}_k)$ and ν_k, which are valid for large values of n. We now have

$$\begin{aligned} k^2 E(\overline{X}_k^2) &= \sum_{i=n-k+1}^{n} \sum_{j=n-k+1}^{n} \mu_{i,j:n} \\ &= 2 \sum_{i=n-k+1}^{n-1} \sum_{j=i+1}^{n} \mu_{i,j:n} + \sum_{i=n-k+1}^{n} \mu_{i:n}^{(2)} \\ &= 2 \sum_{i=n-k+1}^{n-1} \left\{ \sum_{j=i+1}^{n} \mu_{j:n}^{(2)} - (n-i) \right\} + \sum_{i=n-k+1}^{n} \mu_{i:n}^{(2)} \end{aligned} \qquad (2.59)$$

on using equation (2.55). Now upon rearranging the terms in (2.59) and simplifying, we get

$$k^2 E(\overline{X}_k^2) = \sum_{i=n-k+1}^{n} (2i-2n+2k-1)\, \mu_{i:n}^{(2)} - k(k-1). \qquad (2.60)$$

As a result, the mean and variance of $\overline{X}_k$ could be obtained from the tables of $\mu_{i:n}$ and $\mu_{i:n}^{(2)}$ alone. Making use of the tables of the mean and standard deviation of $X_{i:n}$ prepared by Yamauti (1972), the mean and variance of $\overline{X}_k$ have been tabulated by Joshi and Balakrishnan (1981b) for sample sizes up to fifty. They have also pointed out that the approximation for ν_k given by Burrows (1975) is not satisfactory for small values of k even for n = 50. His approximation for ν_k seems to improve with increasing values of k.

For studying outliers in normal samples, Murphy (1951) has suggested the internally studentized version of the selection differential Δ_k given by

$$D_k = k(\overline{X}_k - \overline{X}_n)/S = k\,\Delta_k/S, \qquad (2.61)$$

where $S^2 = \sum_{i=1}^{n} (X_{i:n} - \overline{X}_n)^2/(n-1)$ is the sample variance. Statistics related to D_k for small values of k have been studied in great detail by several authors, see, for example, Barnett and Lewis (1978) and Hawkins (1979) for detailed information on this topic. By applying a theorem due to Basu (1955), we have D_k and S to be statistically independent and, hence, we have from (2.61)

$$E(D_k^m) = k^m E(\Delta_k^m)/E(S^m) = \left\{ k^m \left[\frac{n-1}{2}\right]^{m/2} \Gamma\left[\frac{n-1}{2}\right] / \Gamma\left[\frac{n-1+m}{2}\right] \right\} E\left[\Delta_k^m\right]. \qquad (2.62)$$

The mean and variance of the statistic D_k could now be obtained from equation (2.62) by using the facts

$$E(\Delta_k) = E(\overline{X}_k) \quad \text{and} \quad E(\Delta_k^2) = \mathrm{Var}(\overline{X}_k) - \frac{1}{n} + \left\{E(\overline{X}_k)\right\}^2.$$

These quantities for the case k=1 have also been tabulated by Borenius (1965) for sample sizes up to 120.

2.6. Results for two related populations

In Section 4, we derived several recurrence relations for both single and product moments of order statistics from symmetric populations. In this section, we consider once again the moments of order statistics in samples drawn from a symmetric population and express them in terms of the moments of order statistics in samples drawn from the population obtained by folding the symmetric population at zero. These results have been derived by Govindarajulu (1963b) and have been made use of by Govindarajulu (1966), Gravel and van Eeden (1981), and Balakrishnan and Kocherlakota (1985) to study the order statistics from double exponential, double quadratic and double Weibull populations, respectively.

To this end, let us denote $X_{1:n} < X_{2:n} < \dots < X_{n:n}$ as the order statistics in a sample of size n drawn from a population symmetric about zero, with probability density function f(x) and cumulative distribution function F(x). Let $Y_{1:n} < Y_{2:n} < \dots < Y_{n:n}$ be the order statistics in a sample of size n drawn from the population having its density function p(x) and distribution function P(x) as

$$p(x) = 2\, f(x) \text{ and } P(x) = 2F(x)-1, \qquad x > 0. \tag{2.63}$$

In other words, the distribution of Y's is obtained by folding the distribution of the X's at zero. Further, let us denote the single moments by

$$\mu_{i:n}^{(k)} = E\left[X_{i:n}^k\right] \text{ and } \nu_{i:n}^{(k)} = E\left[Y_{i:n}^k\right], \qquad 1 \le i \le n,\ k > 0,$$

and the product moments by

$$\mu_{i,j:n} = E\left[X_{i:n}X_{j:n}\right] \text{ and } \nu_{i,j:n} = E\left[Y_{i:n}Y_{j:n}\right], \qquad 1 \le i < j \le n.$$

Then Govindarajulu (1963b) has essentially derived some formulae expressing the moments $\mu_{i:n}^{(k)}$ and $\mu_{i,j:n}$ in terms of $\nu_{i:n}^{(k)}$ and $\nu_{i,j:n}$.

Relation 2.32: For $1 \le i \le n$ and $k \ge 1$,

$$2^n \mu_{i:n}^{(k)} = \sum_{r=0}^{i-1} \binom{n}{r} \nu_{i-r:n-r}^{(k)} + (-1)^k \sum_{r=i}^{n} \binom{n}{r} \nu_{r-i+1:r}^{(k)}. \tag{2.64}$$

Proof. From equation (2.2) we have for $1 \le i \le n$ and $k \ge 1$

$$\begin{aligned}\mu_{i:n}^{(k)} &= \{B(i,n-i+1)\}^{-1} \int_{-\infty}^{\infty} x^k \{F(x)\}^{i-1} \{1-F(x)\}^{n-i} f(x)\,dx \\ &= \mathcal{I}_1 + \mathcal{I}_2,\end{aligned} \tag{2.65}$$

where

$$\begin{aligned}\mathcal{I}_1 &= \{B(i,n-i+1)\}^{-1} \int_0^{\infty} x^k \{F(x)\}^{i-1} \{1-F(x)\}^{n-i} f(x)\,dx \\ &= \{B(i,n-i+1)\}^{-1} 2^{-n} \int_0^{\infty} x^k \{1+P(x)\}^{i-1} \{1-P(x)\}^{n-i} p(x)\,dx \\ &= \{B(i,n-i+1)\}^{-1} 2^{-n} \sum_{r=0}^{i-1} \binom{i-1}{r} \int_0^{\infty} x^k \{P(x)\}^{i-r-1} \{1-P(x)\}^{n-i} p(x)\,dx \\ &= 2^{-n} \sum_{r=0}^{i-1} \binom{n}{r} \nu_{i-r:n-r}^{(k)}\end{aligned}$$

and

$$\begin{aligned}\mathcal{I}_2 &= \{B(i,n-i+1)\}^{-1} \int_{-\infty}^{0} x^k \{F(x)\}^{i-1} \{1-F(x)\}^{n-i} f(x)\,dx \\ &= \{B(i,n-i+1)\}^{-1} (-1)^k 2^{-n} \int_0^{\infty} x^k \{1+P(x)\}^{n-i} \{1-P(x)\}^{i-1} p(x)\,dx\end{aligned}$$

$$= \{B(i,n-i+1)\}^{-1} (-1)^k 2^{-n} \sum_{r=0}^{n-i} \binom{n-i}{r} \int_0^\infty x^k \{P(x)\}^r \{1-P(x)\}^{i-1} p(x)\, dx$$

$$= (-1)^k 2^{-n} \sum_{r=i}^{n} \binom{n}{r} \nu^{(k)}_{r-i+1:r}.$$

Equation (2.64) follows upon substituting the above expressions for $\mathcal{I}_1$ and $\mathcal{I}_2$ in equation (2.65).

Relation 2.33: For $1 \le i < j \le n$,

$$2^n \mu_{i,j:n} = \sum_{r=0}^{i-1} \binom{n}{r} \nu_{i-r,j-r:n-r} + \sum_{r=j}^{n} \binom{n}{r} \nu_{r-j+1,r-i+1:r}$$
$$- \sum_{r=i}^{j-1} \binom{n}{r} \nu_{r-i+1:r}\, \nu_{j-r:n-r}. \tag{2.66}$$

Proof. From equation (2.15) we have for $1 \le i < j \le n$

$$\mu_{i,j:n} = C \int_{-\infty}^{\infty} \int_{-\infty}^{y} xy \{F(x)\}^{i-1} \{F(y)-F(x)\}^{j-i-1} \{1-F(y)\}^{n-j} f(x) f(y)\, dx\, dy$$
$$= \mathcal{I}_1 + \mathcal{I}_2 + \mathcal{I}_3, \tag{2.67}$$

where

$$C = \{B(i,j-i,n-j+1)\}^{-1} = n!/(i-1)!(j-i-1)!(n-j)!,$$

$$\mathcal{I}_1 = C \int_0^\infty \int_0^y xy \{F(x)\}^{i-1} \{F(y)-F(x)\}^{j-i-1} \{1-F(y)\}^{n-j} f(x) f(y)\, dx\, dy$$

$$= C\, 2^{-n} \int_0^\infty \int_0^y xy \{1+P(x)\}^{i-1} \{P(y)-P(x)\}^{j-i-1} \{1-P(y)\}^{n-j} p(x) p(y)\, dx\, dy$$

$$= C\, 2^{-n} \sum_{r=0}^{i-1} \binom{i-1}{r} \int_0^\infty \int_0^y xy \{P(x)\}^{i-r-1} \{P(y)-P(x)\}^{j-i-1} \{1-P(y)\}^{n-j} p(x) p(y)\, dx\, dy$$

$$= 2^{-n} \sum_{r=0}^{i-1} \binom{n}{r} \nu_{i-r,j-r:n-r},$$

$$\mathcal{I}_2 = C \int_{-\infty}^{0} \int_{-\infty}^{y} xy \{F(x)\}^{i-1} \{F(y)-F(x)\}^{j-i-1} \{1-F(y)\}^{n-j} f(x) f(y)\, dx\, dy$$

$$= C \int_0^\infty \int_0^v uv \{F(u)\}^{n-j} \{F(v)-F(u)\}^{j-i-1} \{1-F(v)\}^{i-1} f(u) f(v)\, du\, dv$$

$$= C\, 2^{-n} \int_0^\infty \int_0^v uv \{1+P(u)\}^{n-j} \{P(v)-P(u)\}^{j-i-1} \{1-P(v)\}^{i-1} p(u) p(v)\, du\, dv$$

$$= C\, 2^{-n} \sum_{r=0}^{n-j} \binom{n-j}{r} \int_0^\infty \int_0^v uv \{P(u)\}^r \{P(v)-P(u)\}^{j-i-1} \{1-P(v)\}^{i-1} p(u) p(v)\, du\, dv$$

$$= 2^{-n} \sum_{r=j}^{n} \binom{n}{r} \nu_{r-j+1,r-i+1:r}$$

and

$$\mathcal{J}_3 = C \int_0^\infty \int_{-\infty}^0 xy\ \{F(x)\}^{i-1}\ \{F(y)-F(x)\}^{j-i-1}\ \{1-F(y)\}^{n-j}\ f(x)\ f(y)\ dx\ dy$$

$$= -C \int_0^\infty \int_0^\infty xy\ \{1-F(x)\}^{i-1}\ \{F(x)+F(y)-1\}^{j-i-1}\ \{1-F(y)\}^{n-j}\ f(x)\ f(y)\ dx\ dy$$

$$= -C\ 2^{-n} \int_0^\infty \int_0^\infty xy\ \{1-P(x)\}^{i-1}\ \{P(x)+P(y)\}^{j-i-1}\ \{1-P(y)\}^{n-j}\ p(x)\ p(y)\ dx\ dy$$

$$= -C\ 2^{-n} \sum_{r=0}^{j-i-1} \binom{j-i-1}{r} \int_0^\infty x\ \{P(x)\}^r\ \{1-P(x)\}^{i-1}\ p(x)\ dx \int_0^\infty y\ \{P(y)\}^{j-i-r-1}\ \{1-P(y)\}^{n-j}\ p(y)\ dy$$

$$= -2^{-n} \sum_{r=i}^{j-1} \binom{n}{r} \nu_{r-i+1:r}\ \nu_{j-r:n-r}.$$

Equation (2.66) is obtained by substituting the above expressions for $\mathcal{J}_1$, $\mathcal{J}_2$ and $\mathcal{J}_3$ in equation (2.67).

Thus, for example, making use of Relations 2.32 and 2.33 we could obtain the means, variances and covariances of order statistics from a symmetric distribution, given these values for its folded distribution; see, Govindarajulu (1966), Gravel and van Eeden (1981), and Balakrishnan and Kocherlakota (1985). Note that while using Relations 2.32 and 2.33 the error could arise due to approximations either in the coefficients or in the moments of order statistics from the folded distribution (the ν's). Since the coefficients occurring in equations (2.64) and (2.66) are simple binomial coefficients which are integral and could be evaluated exactly at least for small values of n, we may assume that the error involved due to the approximations in the coefficients is zero. Now if ϵ is the error involved in approximating each of the ν's, then the maximum cumulative rounding error in computing $\mu_{i:n}^{(k)}$ $(1 \le i \le n)$ by means of Relation 2.32 is given by

$$2^{-n}\left\{ \sum_{r=0}^{i-1} \binom{n}{r} + \sum_{r=i}^{n} \binom{n}{r} \right\} \epsilon = 2^{-n} \sum_{r=0}^{n} \binom{n}{r} \epsilon = \epsilon.$$

That is, the maximum error involved in computing numerically $\mu_{i:n}^{(k)}$ $(1 \le i \le n)$ using the moments of the order statistics from the folded distribution, is at most ϵ where ϵ is the maximum error involved in approximating each ν. Similarly, the cumulative round off error involved in the numerical evaluation of the product moments $\mu_{i,j:n}$ $(1 \le i < j \le n)$ by means of Relation 2.33 has also been studied and a bound on the maximum error has been obtained by Govindarajulu (1963b).

2.7. Results for exchangeable variates

Several recurrence relations and identities for moments of order statistics have been presented in Sections 1 and 2. All these results have been derived for order statistics obtained from a sample of independent and indentically distributed variables. Many of these results continue to hold when the order statistics arise from a sample of exchangeable variables.

Let $X_1, X_2, \ldots, X_n$ be n exchangeable random variables with joint cumulative distribution function

$F(x_1,x_2,\dots,x_n)$; that is, $F(x_1,x_2,\dots,x_n)$ is symmetrical in $x_1,x_2,\dots,x_n$. As before, let $X_{1:n} \le X_{2:n} \le \dots \le X_{n:n}$ be the order statistics obtained from the above set of n exchangeable random variables. Then, David and Joshi (1968) have established Relation 2.2 by a simple argument which is as follows. Drop one order statistic $X_{r:n}$ $(1 \le r \le n)$ at random from the set of n order statistics. Then it is easy to note that the remaining n–1 are simply the order statistics obtained from a sample of n–1 exchangeable variables. Now, with probability i/n, if one of the smallest i order statistics is dropped, the i'th order statistic in the sample of n–1 would be the (i+1)'th order statistic in the sample of n. Similarly, with probability (n–i)/n, if one of the largest n–i order statistics is dropped, the i'th order statistic in the sample of n–1 would be the i'th order statistic in the sample of n. As a result, we obtain the relation

$$\frac{i}{n} \Pr(X_{i+1:n} \le x) + \frac{n-i}{n} \Pr(X_{i:n} \le x) = \Pr(X_{i:n-1} \le x) \tag{2.68}$$

for all x. Then by a repeated application of equation (2.68) we obtain the relations

$$\Pr(X_{i:n} \le x) = \sum_{j=i}^{n} (-1)^{j-i} \binom{n}{j} \binom{j-1}{i-1} \Pr(X_{j:j} \le x) \tag{2.69}$$

and

$$\Pr(X_{i:n} \le x) = \sum_{j=n-i+1}^{n} (-1)^{j-n+i-1} \binom{n}{j} \binom{j-1}{n-i} \Pr(X_{1:j} \le x). \tag{2.70}$$

Even though these results have been given for the cumulative distribution functions of order statistics, it is also easy to obtain similar relations for the single moments. In this respect equations (2.68)–(2.70) are generalizations of the results in Section 1 from the case of independent variables to that of exchangeable random variables. For some extensions of these results, one may refer to Maurer and Margolin (1976), Galambos (1978), and David (1981). Making use of equations (2.68)–(2.70), Balakrishnan (1987a) has derived some additional identities for this case.

For the case when the variables X_i $(1 \le i \le n)$ have a standard multivariate normal distribution with common correlation coefficient ρ, Gupta (1963) has tabulated the probability $\Pr(X_{n:n} \le x)$ for various choices of $\rho > 0$ and sample sizes up to 12. With the help of equation (2.68), David and Joshi (1968) have obtained upper percentage points of $X_{n-1:n}$ from Gupta's (1963) tables; also see Young (1967). For some pertinent work on this line, interested readers may refer to Bland and Owen (1966), Greig (1967), Afonja (1972), Hoffman and Saw (1975), Rawlings (1976), and Hill (1976).

The method of proof of David and Joshi (1968) could also be easily applied to establish some relations for the joint distributions of two order statistics obtained from a set of n exchangeable variables. For example, by dropping one order statistic $X_{r:n}$ $(1 \le r \le n)$ at random from the set of n order statistics and proceeding as before, we obtain the relation

$$\frac{i}{n} \Pr(X_{i+1:n} \le x, X_{j+1:n} \le y) + \frac{j-i}{n} \Pr(X_{i:n} \le x, X_{j+1:n} \le y) + \frac{n-j}{n} \Pr(X_{i:n} \le x, X_{j:n} \le y) = \Pr(X_{i:n-1} \le x, X_{j:n-1} \le y),$$

which, of course, generalizes Relation 2.13 to the case when the order statistics arise from a set of exchangeable random variables.

Exercises

1. (Galton (1902); Pearson (1902)). For a random sample of size n from a population with distribution function F(x), show that for i = 1,2,...,n−1

$$\mu_{i+1:n} - \mu_{i:n} = \binom{n}{i} \int_{-\infty}^{\infty} \{F(x)\}^i \{1-F(x)\}^{n-i}\, dx.$$

2. (Sillitto (1964)). For $1 \le i \le m < n$, show that

$$\binom{n}{m} \mu_{i:m}^{(k)} = \sum_{r=0}^{n-m} \binom{i+r-1}{r} \binom{n-i-r}{m-i} \mu_{i+r:n}^{(k)}.$$

3. (Downton (1966)). For any arbitrary distribution, show that for $j+j' \le n-1$

$$\sum_{i=j+1}^{n-j'} (i-1)^{(j)} (n-i)^{(j')} \mu_{i:n}^{(k)} = j!\, j'! \binom{n}{j+j'+1} \mu_{j+1:j+j'+1}^{(k)},$$

and for $j+j' \le n-2$

$$\sum_{i=j+1}^{n-1-j'} \sum_{i'=i+1}^{n-j'} (i-1)^{(j)}(n-i')^{(j')}\mu_{i,j:n} = j!j'! \binom{n}{j+j'+2} \mu_{j+1,j+2:j+j'+2},$$

where $n^{(i)} = n(n-1)(n-2)...(n-i+1)$.

4. (Joshi (1973)). For $n \ge 2$, show that

$$\sum_{i=1}^{n} \frac{1}{i} \mu_{i:n}^{(k)} = \sum_{i=1}^{n} \frac{1}{i} \mu_{1:i}^{(k)}$$

and

$$\sum_{i=1}^{n} \frac{1}{(n-i+1)} \mu_{i:n}^{(k)} = \sum_{i=1}^{n} \frac{1}{i} \mu_{i:i}^{(k)}.$$

5. (Balakrishnan and Malik (1985)). By denoting $C_r = (n+1)(n+2)...(n+r-1)$ for $r \ge 2$, prove that the following identities hold for any arbitrary distribution:

for $r \ge 2$,

$$\sum_{i=1}^{n} \mu_{i:n}^{(k)}/\{i(i+1)...(i+r-1)\} = \frac{1}{C_r} \sum_{i=1}^{n} \binom{i+r-2}{r-1} \mu_{1:i}^{(k)}/i;$$

for $r \ge 2$,

$$\sum_{i=1}^{n} \mu_{i:n}^{(k)}/\{(n-i+1)(n-i+2)...(n-i+r)\} = \frac{1}{C_r} \sum_{i=1}^{n} \binom{i+r-2}{r-1} \mu_{i:i}^{(k)}/i;$$

for $r \ge 1$,

$$\sum_{i=1}^{n} \mu_{i:n}^{(k)}/\{i(i+1)...(i+r-1)(n-i+1)(n-i+2)...(n-i+r)\} = \frac{1}{C_{2r}} \sum_{i=1}^{n} \begin{bmatrix} i+2r-2 \\ r-1 \end{bmatrix} \{\mu_{1:i}^{(k)} + \mu_{i:i}^{(k)}\}/i;$$

and for r,s ≥ 1,

$$\sum_{i=1}^{n} \mu_{i:n}^{(k)}/\{i(i+1)...(i+r-1)(n-i+1)(n-i+2)...(n-i+s)\}$$

$$= \frac{1}{C_{r+s}} \sum_{i=1}^{n} \left\{ \begin{bmatrix} i+r+s-2 \\ r-1 \end{bmatrix} \mu_{1:i}^{(k)} + \begin{bmatrix} i+r+s-2 \\ s-1 \end{bmatrix} \mu_{i:i}^{(k)} \right\}/i.$$

6. (Sillitto (1951); Cadwell (1953)). Denoting $\omega_n = \mu_{n:n} - \mu_{1:n}$ $(n \geq 2)$, show that for $n \geq 3$

$$n\,\omega_{n-1} - (n-1)\,\omega_n = \mu_{n-1:n} - \mu_{2:n}$$

and

$$\omega_n = \omega_{n-1} + \frac{1}{n}(\chi_{1:n} + \chi_{n-1:n})$$

(recall $\chi_{i:n} = \mu_{i+1:n} - \mu_{i:n}$).

7. (Romanovsky (1933); Sillitto (1951)). For any arbitrary continuous distribution, show that

$$\begin{bmatrix} n \\ i \end{bmatrix} \sum_{r=0}^{i} (-1)^{r+1} \begin{bmatrix} i \\ r \end{bmatrix} \omega_{n-i+r} = \chi_{i:n} + \chi_{n-i:n}, \quad 1 \leq i \leq n-1,$$

$$\{1 - (-1)^n\}\,\omega_n = \sum_{j=2}^{n-1} (-1)^j \begin{bmatrix} n \\ j \end{bmatrix} \omega_j, \quad n \geq 3,$$

and, consequently,

$$2\,\omega_n = \sum_{j=2}^{n-1} (-1)^j \begin{bmatrix} n \\ j \end{bmatrix} \omega_j$$

for odd values of n .

8. (Sillitto (1951)). Show for any arbitrary distribution,

$$n\,\chi_{i-1:n-1} - (n-i+1)\,\chi_{i-1:n} = i\,\chi_{i:n}, \quad 2 \leq i \leq n-1,$$

and for $m \leq i-1$

$$\chi_{i:n} = \frac{(n)^{(m)}}{(i)^{(m)}} \sum_{j=0}^{m} (-1)^j \begin{bmatrix} m \\ j \end{bmatrix} \frac{(n-i+j)^{(j)}}{(n-m+j)^{(j)}} \chi_{i-m:n-m+j},$$

where, as before, $(n)^{(m)} = n(n-1)(n-2)...(n-m+1)$.

9. (Govindarajulu (1963a)). For i = 0,1,2,...,[(n−2)/2], prove for any arbitrary distribution

$$i\,\omega_{i,n} + (n-i)\,\omega_{i-1,n} = n\,\omega_{i-1,n-1},$$

where $\omega_{i,n}$ is the expected value of the i'th quasi–range in a sample of size n, viz., $\mu_{n-i:n} - \mu_{i+1:n}$.

10. (Balakrishnan (1986)). By writing $\mu_{i:n}^{(k)}$ for a non–negative integer valued population as

$$\mu_{i:n}^{(k)} = \sum_{x=0}^{\infty} x^k f_{i:n}(x)$$

where

$$f_{i:n}(x) = \{B(i,n-i+1)\}^{-1} \int_{P(x-1)}^{P(x)} t^{i-1} (1-t)^{n-i} dt,$$

obtain alternative derivations of the results given in Exercises 2–9 for discrete populations.

11. (Govindarajulu (1968)). Show that for a population with cumulative distribution function F(x)

$$E\left\{F(X_{i:n}) X_{j:n}\right\} = \frac{i}{n+1} \mu_{j+1:n+1}, \quad 1 \le i \le j \le n$$

and

$$E\left\{X_{i:n} F(X_{j:n})\right\} = \mu_{i:n} - \frac{n+1-j}{n+1} \mu_{i:n+1}, \quad 1 \le i < j \le n.$$

12. (Tippett (1925); Cox (1954); Siotani (1957); Carlson (1958)). For the sample range $R_n = X_{n:n} - X_{1:n}$, show that for a population

$$E(R_n) = \int_{-\infty}^{\infty} \left[1 - \{F(x)\}^n - \{1-F(x)\}^n\right] dx$$

and

$$Var(R_n) = 2 \int_{-\infty}^{\infty} \int_{-\infty}^{y} \left[1-\{F(y)\}^n - \{1-F(x)\}^n + \{F(y)-F(x)\}^n\right] dx\, dy - \{E(R_n)\}^2;$$

for a non-negative integer valued population, show that

$$E(R_n) = \sum_{x=0}^{\infty} \left[1-\{F(x)\}^n - \{1-F(x)\}^n\right]$$

and

$$Var(R_n) = 2 \sum_{y=0}^{\infty} \sum_{x=0}^{y} \left[1-\{F(y)\}^n-\{1-F(x)\}^n+\{F(y)-F(x)\}^n\right] - E(R_n) - \{E(R_n)\}^2.$$

For an arbitrary distribution, show also that

$$E(R_{2n+1}) = (2n+1) \sum_{i=1}^{n-1} (-1)^{n+i+1} E(R_{n+i+1})/(n+i+1).$$

13. (Joshi and Balakrishnan (1982)). For any population, show that for $1 \le i \le n-2$

$$(n-i)(n-i-1) \sum_{j=0}^{i} \left\{\binom{i}{j} \Big/ \binom{n-2}{j}\right\} \sum_{k=0}^{j} \mu_{n-j-1,n-k:n} = n(n-1)\, \mu_{n-i-1,n-i:n-i}.$$

14. For any distribution symmetric about zero, show that for $1 \le i \le n-1$

$$\sum_{s=i+1}^{n} \mu_{i,s:n} + \sum_{r=1}^{i} \mu_{r,i+1:n} = 0,$$

and, in particular, for even values of n, say n = 2m,

$$\sum_{i=1}^{m} \mu_{i,m+1:2m} = 0.$$

15. (Balakrishnan (1982)). For any arbitrary population, show that for $n \geq 2$

$$\sum_{i=1}^{n-1} \mu_{i,i+1:n} = \sum_{i=1}^{n-1} \binom{n}{i} \mu_{1:n-i}\, \mu_{i:i} - \sum_{i=2}^{n} \binom{n}{i} \mu_{1,i:i}$$

and, in general, for $1 \leq r \leq n-1$

$$\sum_{i=1}^{n-r} \mu_{i,i+r:n} = \sum_{i=1}^{n-r} \sum_{j=1}^{r} (-1)^{j-1} \binom{n}{i+r-j} \binom{n-i-r+j-1}{j-1} \mu_{1:n-i-r+j}\, \mu_{i:i+r-j}$$

$$+ (-1)^{r} \sum_{i=1}^{n-r} \binom{i+r-2}{r-1} \binom{n}{i+r} \mu_{1,i+r:i+r}.$$

For even values of n, then deduce that

$$\mu_{1,n:n} = \sum_{i=1}^{(n-2)/2} (-1)^{i-1} \binom{n}{i} \mu_{1:i}\, \mu_{1:n-i} + \frac{1}{2} (-1)^{(n-2)/2} \binom{n}{n/2} \mu^{2}_{1:n/2}.$$

16. (Khatri (1962)). For a non–negative integer valued population, show that the joint probability mass function of $X_{i:n}$ and $X_{j:n}$ $(1 \leq i < j \leq n)$ can be written as

$$f_{i,j:n}(x,y) = \Pr(X_{i:n} = x,\, X_{j:n} = y)$$

$$= \{B(i,j-i,n-j+1)\}^{-1} \iint u^{i-1} (v-u)^{j-i-1} (1-v)^{n-j}\, du\, dv,$$

where the integration is over the region $u \leq v$, $F(x-1) \leq u \leq F(x)$ and $F(y-1) \leq v \leq F(y)$.

17. (Balakrishnan (1986)). By using the above form for the joint p.m.f. of $X_{i:n}$ and $X_{j:n}$ $(1 \leq i < j \leq n)$, present an alternative derivation of the results of Exercises 13–15 in the discrete case.

18. (Govindarajulu (1963a)). Proceeding exactly on the same lines as in the proof of Relation 2.26, prove the following results for an arbitrary continuous population:

(i) If g is a twice differentiable function such that twice differentiation of g(x) with respect to x and expectation of g(X) with respect to an absolutely continuous distribution are interchangeable, then for $1 \leq i \leq n$

$$E\{g''(X_{i:n})\} = E\{g(X_{i:n}) \sum_{j=1}^{n} f''(X_{j:n})/f(X_{j:n})\}$$

$$+ E\{g(X_{i:n}) \sum\sum_{j \neq k} f'(X_{j:n}) f'(X_{k:n})/f(X_{j:n}) f(X_{k:n})\};$$

(ii) If g and h are differentiable functions such that differentiation of g(x) h(x) with respect to x and expectation of g(X) h(X) with respect to an absolutely continuous distribution are interchangeable, then

$$E\{g'(X_{i:n})\, h(X_{j:n}) + g(X_{i:n})\, h'(X_{j:n})\} = -\sum_{k=1}^{n} E\{g(X_{i:n})\, h(X_{j:n})\, f'(X_{k:n})/f(X_{k:n})\}.$$

19. (McKay (1935)). Let X_i $(i = 1,2,\ldots,n)$ be a random sample of size n from a normal $N(\mu,\sigma^2)$ population. Then show that $\bar{X}$ and $X_{i:n}-\bar{X}$ $(1 \le i \le n)$ are statistically independent. Also show that $\sum_{i=1}^{n} a_i X_{i:n}/S$, with $\sum_{i=1}^{n} a_i = 0$, is independent of S, where S^2 is the sample variance.

20. (David (1973)). Let X_i $(1 \le i \le n)$ be a random sample from a population with pdf f(x) and cdf F(x), and let Y be another independent variable with pdf g(x). Then show that for $1 \le i \le n$,

$$E(X_{i:n} \mid X_{i:n} \le Y < X_{i+1:n}) = \frac{i \int_{-\infty}^{\infty} \int_{x}^{\infty} x\{F(x)\}^{i-1}\, \{1-F(y)\}^{n-i}\, f(x)\, g(y)\, dy\, dx}{\int_{-\infty}^{\infty} \{F(y)\}^{i}\, \{1-F(y)\}^{n-i}\, g(y)\, dy}.$$

21. (Govindarajulu (1963b)). Show that the maximum cumulative rounding error in evaluating $\mu_{i,j:n}$ $(1 \le i < j \le n)$ by using Relation 2.33 in Section 6 is given by

$$\epsilon + \epsilon\, 2^{-(n-1)} \binom{n}{n/2} (j-i)\, \nu_{j-i:j-i},$$

where ϵ is the maximum error involved in approximating any one of the ν's.

22. (Joshi and Balakrishnan (1982)). For an arbitrary population, show that for $n \ge 3$

$$\sum_{i=1}^{n-1} \mu_{i,i+1:n} + \sum_{j=2}^{n} \binom{n}{j} \mu_{1,j:j} = \sum_{j=1}^{n-1} \binom{n}{j} \mu_{j:j}\, \mu_{1:n-j}.$$

23. (Balakrishnan (1987a)). Making use of equations (2.69) and (2.70), show that the results given in Exercises 4 and 5 also hold for the case when the order statistics are obtained from n exchangeable random variables.

24. (Heath and Sudderth (1976); de Finetti (1937)). If $X_1, X_2, X_3, \ldots$ form an exchangeable sequence, that is, the X_i's in every finite subset are exchangeable, then show that there exists a random variable V such that

$$\Pr\{X_1 \le x_1, X_2 \le x_2, \ldots, X_n \le x_n\} = E\left[\prod_{i=1}^{n} \Pr\{X_i \le x_i \mid V\}\right],$$

where the conditional distributions are identical.

25. (Bhattacharyya (1970); Dykstra et al. (1973)). If $X_1, X_2, X_3, \ldots$ form an exchangeable sequence, then show that for all x

$$\Pr(X_{1:n} \le x) \le 1 - \{1-F(x)\}^n$$

and

$$\Pr(X_{n:n} \le x) \ge \{F(x)\}^n;$$

consequently,

$$E(W_n) \le \int_{-\infty}^{\infty} \left[1-\{1-F(x)\}^n - \{F(x)\}^n\right] dx$$

where $W_n = X_{n:n} - X_{1:n}$ is the sample range in a sample of size n.

26. (Balakrishnan (1986)). By using the joint probability mass function of $X_{i:n}$ and $X_{j:n}$ $(1 \le i < j \le n)$ given in Exercise 16, show that for an arbitrary discrete population (x = 0,1,2,...) we could write the product moments $\mu_{i,j:n}$ as

$$\mu_{i,i+1:n} = \mu_{i:n}^{(2)} + \binom{n}{i} \sum_{x=0}^{\infty} x\left[\{F(x)\}^i - \{F(x-1)\}^i\right] \sum_{y=x}^{\infty} \{1-F(y)\}^{n-i}, \qquad 1 \le i \le n-1,$$

and for $1 \le i < j \le n$, $j-i \ge 2$,

$$\mu_{i,j:n} = \sum_{r=0}^{j-i-1} \sum_{s=0}^{r} (-1)^{j-i-1-s} \binom{n}{s} \binom{i-1+r-s}{r-s} \binom{n-i-r-1}{j-i-r-1} \left[\mu_{i+r-s:n-s}^{(2)}\right.$$

$$\left. + \binom{n-s}{i+r-s} \sum_{x=0}^{\infty} x\left\{\{F(x)\}^{i+r-s} - \{F(x-1)\}^{i+r-s}\right\} \sum_{y=x}^{\infty} \{1-F(y)\}^{n-i-r}\right].$$

27. (Margolin and Winokur (1967); Balakrishnan (1986)). For the geometric distribution with pmf

$$f(x) = \Pr(X = x) = p\, q^{x-1}, \qquad x = 1,2,\ldots,\ q = 1-p,$$

show that for $1 \le i \le n$

$$\mu_{i:n} = \sum_{j=n-i+1}^{n} (-1)^{j-n+i-1} \binom{j-1}{n-i} \binom{n}{j} \frac{1}{(1-q^j)},$$

$$\mu_{i:n}^{(2)} = \sum_{j=n-i+1}^{n} (-1)^{j-n+i-1} \binom{j-1}{n-i} \binom{n}{j} \frac{(1+q^j)}{(1-q^j)^2},$$

and for $1 \le i \le n-1$

$$\mu_{i,i+1:n} = \mu_{i:n}^{(2)} - \binom{n}{i} \left\{\frac{q^{n-i}}{1-q^{n-i}}\right\} \sum_{j=0}^{i} (-1)^{i-j} \binom{i}{j} \frac{(1-q^{i-j})}{(1-q^{n-j})^2}.$$

28. (Barnett (1966)). For the Cauchy distribution with pdf

$$f(x) = \frac{1}{\pi(1+x^2)}, \qquad -\infty < x < \infty,$$

show that for $3 \le i \le n-2$,

$$\mu_{i:n}^{(2)} = \frac{n}{\pi} (\mu_{i:n-1} - \mu_{i-1:n-1}) - 1,$$

and, in general, for $k+1 \le i \le n-k$,

$$\mu_{i:n}^{(k)} = \frac{n}{(k-1)\pi} \left\{\mu_{i:n-1}^{(k-1)} - \mu_{i-1:n-1}^{(k-1)}\right\} - \mu_{i:n}^{(k-2)}.$$

29. (Joshi (1978, 1982)). For the standard exponential distribution with pdf

$$f(x) = e^{-x}, \qquad 0 \le x < \infty,$$

show that the moments of order statistics satisfy the following recurrence relations:

$$\mu_{1:n}^{(k)} = \frac{k}{n}\mu_{1:n}^{(k-1)}, \quad n \ge 1,\ k \ge 1,$$

$$\mu_{i:n}^{(k)} = \mu_{i-1:n-1}^{(k)} + \frac{k}{n}\mu_{i:n}^{(k-1)}, \quad 2 \le i \le n,\ k \ge 1,$$

$$\mu_{i,i+1:n} = \mu_{i:n}^{(2)} + \frac{1}{n-i}\mu_{i:n}, \quad 1 \le i \le n-1,$$

and

$$\mu_{i,j:n} = \mu_{i,j-1:n} + \frac{1}{n-j+1}\mu_{i:n}, \quad 1 \le i < j \le n,\ j-i \ge 2.$$

Also refer to Joshi (1979a) and Balakrishnan and Joshi (1984) for some similar results on truncated exponential populations.

30. (Shah (1966, 1970)). For the logistic distribution with pdf

$$f(x) = e^{-x}/(1+e^{-x})^2, \quad -\infty < x < \infty,$$

show that

$$\mu_{i+1:n+1}^{(k)} = \mu_{i:n}^{(k)} + \frac{k}{i}\mu_{i:n}^{(k-1)}, \quad 1 \le i \le n,\ k \ge 1,$$

$$\mu_{i,i+1:n+1} = \frac{n+1}{n-i+1}\left[\mu_{i,i+1:n} - \frac{i}{n+1}\mu_{i+1:n+1}^{(2)} - \frac{1}{n-i}\mu_{i:n}\right], \quad 1 \le i \le n-1,$$

and

$$\mu_{i,j:n+1} = \frac{n+1}{n-j+2}\left[\mu_{i,j:n} - \mu_{i,j-1:n} + \frac{n-j+2}{n+1}\mu_{i,j-1:n+1} - \frac{1}{n-j+1}\mu_{i:n}\right], \quad 1 \le i < j \le n,\ j-i \ge 2.$$

Also refer to Balakrishnan and Joshi (1983) and Balakrishnan and Kocherlakota (1986) for some similar results on truncated logistic populations.

31. (Dyer and Whisenand (1973a,b); Balakrishnan et al. (1988)). By considering the Rayleigh distribution with pdf

$$f(x) = x\, e^{-x^2/2}, \quad 0 \le x < \infty,$$

derive the following recurrence relations:

$$\mu_{1:n}^{(k+2)} = \frac{k+2}{n}\mu_{1:n}^{(k)}, \quad n \ge 1,\ k \ge 0,$$

$$\mu_{i:n}^{(k+2)} = \mu_{i-1:n}^{(k+2)} + \frac{k+2}{n-i+1}\mu_{i:n}^{(k)}, \quad 1 \le i \le n,\ k \ge 0,$$

$$\mu_{i,i+1:n}^{(k,\ell+2)} = \mu_{i:n}^{(k+\ell+2)} + \frac{\ell+2}{n-i}\mu_{i,i+1:n}^{(k,\ell)}, \quad 1 \le i \le n-1,\ k,\ell \ge 0,$$

$$\mu_{i,j:n}^{(k,\ell+2)} = \mu_{i,j-1:n}^{(k,\ell+2)} + \frac{\ell+2}{n-j+1}\mu_{i,j:n}^{(k,\ell)}, \quad 1 \le i < j \le n,\ j-i \ge 2,\ k,\ell \ge 0,$$

$$\mu_{1,2:n}^{(k+2,\ell)} = (k+2)\,\mu_{1,2:n}^{(k,\ell)} - (n-1)\,\mu_{1:n}^{(k+\ell+2)}, \quad n \ge 2,\ k,\ell \ge 0,$$

$$\mu_{i,i+1:n}^{(k+2,\ell)} = \frac{k+2}{i}\mu_{i,i+1:n}^{(k,\ell)} - \frac{1}{i}\left\{(n-i)\mu_{i:n}^{(k+\ell+2)} - n\,\mu_{i-1,i:n-1}^{(k+2,\ell)}\right\}, \quad 2 \le i \le n-1,\ k,\ell \ge 0,$$

and

$$\mu_{i,j:n}^{(k+2,\ell)} = \mu_{i+1,j:n}^{(k+2,\ell)} - \frac{n}{i}\left\{\mu_{i,j-1:n-1}^{(k+2,\ell)} - \mu_{i-1,j-1:n-1}^{(k+2,\ell)}\right\} + \frac{k+2}{i}\mu_{i,j:n}^{(k,\ell)},$$

$$1 \le i < j \le n,\ j-i \ge 2,\ k,\ell \ge 0,$$

where $\mu_{i,j:n}^{(k,\ell)}$ denotes $E(X_{i:n}^k\, X_{j:n}^\ell)$. Also see Balakrishnan and Malik (1986b) for generalizations of these results to the linear–exponential population with increasing hazard rate.

32. (Krishnaiah and Rizvi (1966)). For an arbitrary distribution and a specified function h such that

E(h(X)) exists, show that the following relations hold:
for $1 \le i \le n$

$$E\{h(X_{i:n})\} = \binom{n}{i} \sum_{j=0}^{r} (-1)^j \frac{i}{i-r} \frac{\binom{r}{j}}{\binom{n-r+j}{i-r}} E\{h(X_{i-r:n-r+j})\}, \quad 0 \le r \le i-1,$$

and

$$E\{h(X_{i:n})\} = \binom{n}{i} \sum_{j=0}^{r} (-1)^j \frac{i}{i+j} \frac{\binom{r}{j}}{\binom{n-r+j}{i+j}} E\{h(X_{i+j:n-r+j})\}, \quad 0 \le r \le n-i.$$

33. (Mustafi (1969)). Let F(x,y) be a bivariate distribution function with marginals $F_1(x)$ and $F_2(y)$. Consider n independent bivariate observations from this population and let $X_{1:n} < X_{2:n} < ... < X_{n:n}$ and $Y_{1:n} < Y_{2:n} < ... < Y_{n:n}$ be the two sets of order statistics corresponding to X and Y. Then denoting

$$G_{i,j:n}(x,y) = Pr(X_{i:n} \le x, Y_{j:n} \le y), \quad i,j = 1,2,...,n,$$

and the difference operators by

$$\Delta_1 G_{i,j:n}(x,y) = G_{i+1,j:n}(x,y) - G_{i,j:n}(x,y),$$
$$\Delta_2 G_{i,j:n}(x,y) = G_{i,j+1:n}(x,y) - G_{i,j:n}(x,y)$$

and

$$\Delta_{1,2} G_{i,j:n}(x,y) = \Delta_1\Delta_2 G_{i,j:n}(x,y)$$
$$= G_{i+1,j+1:n}(x,y) - G_{i,j+1:n}(x,y) - G_{i+1,j:n}(x,y) + G_{i,j:n}(x,y),$$

show that

$$\frac{i}{n} \Delta_1 G_{i,j:n}(x,y) = F_1(x)\, \Delta_1 G_{i-1,j:n-1}(x,y) - F(x,y)\, \Delta_{1,2} G_{i-1,j-1:n-1}(x,y)$$

for $2 \le i,\ j < n-1$.

34. (Balakrishnan (1985); Balakrishnan and Puthenpura (1986)). For the half logistic distribution with pdf

$$f(x) = 2\, e^{-x}/(1+e^{-x})^2, \quad 0 \le x < \infty,$$

derive the following recurrence relations:

$$\mu_{1:n+1}^{(k+1)} = 2\left[\mu_{1:n}^{(k+1)} - \frac{k+1}{n} \mu_{1:n}^{(k)}\right], \quad n \ge 1, k \ge 0,$$

$$\mu_{2:n+1}^{(k+1)} = \frac{(n+1)(k+1)}{n} \mu_{1:n}^{(k)} - \frac{n-1}{2} \mu_{1:n+1}^{(k+1)}, \quad n \ge 1, k \ge 0,$$

$$\mu_{i+1:n+1}^{(k+1)} = \frac{1}{i}\left[\frac{(n+1)(k+1)}{n-i+1} \mu_{i:n}^{(k)} + \frac{n+1}{2} \mu_{i-1:n}^{(k+1)} - \frac{n-2i+1}{2} \mu_{i:n+1}^{(k+1)}\right], \quad 2 \le i \le n, k \ge 0,$$

$$\mu_{i,i+1:n+1} = \mu_{i:n+1}^{(2)} + \frac{2(n+1)}{n-i+1}\left[\mu_{i,i+1:n} - \mu_{i:n}^{(2)} - \frac{1}{n-i} \mu_{i:n}\right], \quad 1 \le i \le n-1,$$

$$\mu_{2,3:n+1} = \mu_{3:n+1}^{(2)} + (n+1)\left\{\mu_{2:n} - \frac{n}{2} \mu_{1:n-1}^{(2)}\right\}, \quad n \ge 2,$$

$$\mu_{i+1,i+2:n+1} = \mu_{i+2:n+1}^{(2)} + \frac{n+1}{i(i+1)}\left[2\mu_{i+1:n} + n\left\{\mu_{i-1,i:n-1} - \mu_{i:n-1}^{(2)}\right\}\right], \quad 2 \le i \le n-1,$$

$$\mu_{i,j:n+1} = \mu_{i,j-1:n+1} + \frac{2(n+1)}{n-j+2}\left[\mu_{i,j:n} - \mu_{i,j-1:n} - \frac{1}{n-j+1} \mu_{i:n}\right], \quad 1 \le i \le n-2, j-i \ge 2,$$

$$\mu_{2,j+1:n+1} = \mu_{3,j+1:n+1} + (n+1)\left[\mu_{j:n} - \frac{n}{2} \mu_{1,j-1:n-1}\right], \quad 3 \le j \le n,$$

and

$$\mu_{i+1,j+1:n+1} = \mu_{i+2,j+1:n+1} + \frac{n+1}{i(i+1)}\left[2\mu_{j:n} - n\left\{\mu_{i,j-1:n-1} - \mu_{i-1,j-1:n-1}\right\}\right],$$
$$2 \le i \le n-2,\ j-i \ge 2.$$

35. (Ragab and Green (1984); Balakrishnan et al. (1988); Balakrishnan and Malik (1987)). For the log-logistic distribution with pdf

$$f(x) = \beta\, x^{\beta-1}/(1+x^{\beta})^2, \quad x \ge 0,\ \beta > 1,$$

making use of the result by Sen (1959) show that $\mu_{i:n}^{(k)}$ exists for all i ($1 \le i \le n$) only if $k < \beta$. Also in this case, derive the following recurrence relations:

$$\mu_{1:n+1}^{(k)} = \left(1 - \frac{k}{n\beta}\right) \mu_{1:n}^{(k)}, \quad n \ge 1,\ k \ge 1,$$

$$\mu_{i+1:n+1}^{(k)} = \left(1 + \frac{k}{i\beta}\right) \mu_{i:n}^{(k)}, \quad 1 \le i \le n,\ k \ge 1,$$

$$\mu_{i,i+1:n+1} = \frac{1}{n-i+1}\left[(n+1)\left\{1 - \frac{1}{(n-i)\beta}\right\} \mu_{i,i+1:n} - i\, \mu_{i+1:n+1}^{(2)}\right], \quad 1 \le i \le n-1,$$

$$\mu_{i+1,i+2:n+1} = \frac{1}{i+1}\left[(n+1)\left(1 + \frac{1}{i\beta}\right)\mu_{i,i+1:n} - (n-i)\mu_{i+1:n+1}^{(2)}\right], \quad 1 \le i \le n-1,$$

$$\mu_{i,j:n+1} = \mu_{i,j-1:n+1} + \frac{n+1}{n-j+2}\left[\left\{1 - \frac{1}{(n-j+1)\beta}\right\} \mu_{i,j:n} - \mu_{i,j-1:n}\right], \quad 1 \le i < j \le n,\ j-i \ge 2,$$

and

$$\mu_{i+1,j+1:n+1} = \mu_{i+2,j+1:n+1} + \frac{n+1}{i+1}\left[\left(1 + \frac{1}{i\beta}\right) \mu_{i,j:n} - \mu_{i+1,j:n}\right], \quad 1 \le i < j \le n,\ j-i \ge 2.$$

36. (Gupta (1960); Joshi (1979b)). For the gamma distribution with density function

$$f(x) = \frac{1}{\Gamma(\lambda)}\, e^{-x}\, x^{\lambda-1}, \quad x > 0,\ \lambda = 1,2,\ldots,$$

derive the following relations:

$$\mu_{i:n}^{(k)} = \frac{n!}{(i-1)!\,(n-i)!} \sum_{j=0}^{i-1} (-1)^j \binom{i-1}{j} \frac{1}{n-i+j+1}\, \mu_{1:n-i+j+1}^{(k)};$$

$$\mu_{1:n}^{(k)} = \frac{k}{n}\,\Gamma(\lambda) \sum_{j=0}^{\lambda-1} \mu_{1:n}^{(k+j-\lambda)}/j!, \quad k \ge 1,$$

and

$$\mu_{i:n}^{(k)} = \mu_{i-1:n-1}^{(k)} + \frac{k}{n}\,\Gamma(\lambda) \sum_{j=0}^{\lambda-1} \mu_{i:n}^{(k+j-\lambda)}/j!, \quad 2 \le i \le n,\ k \ge 1.$$

Note that the first relation can also be obtained from Exercise 32 by setting $i = r-1$. The last two relations enable us to obtain certain negative moments of order statistics.

37. Let $X_{1:n} \le X_{2:n} \le \ldots \le X_{n:n}$ be the order statistics obtained from a random sample of size n from an arbitrary population. Show that

$$\mu_{i:n}^{(k)} = \sum_{r=0}^{i-1} \sum_{s=0}^{r} (-1)^s \binom{n}{r} \binom{r}{s}\, \mu_{1:n-r+s}^{(k)}$$

and

$$\mu_{i:n}^{(k)} = \sum_{r=i}^{n} \sum_{s=0,1}^{r} (-1)^{s+1} \begin{bmatrix} n \\ r \end{bmatrix} \begin{bmatrix} r \\ s \end{bmatrix} \mu_{1:n-r+s}^{(k)},$$

where the summation from s = 0,1 to r denotes a sum from 0 to r (r < n) and from 1 to r (r = n). Young (1970) proved this result assuming that the variables are nonnegative integer valued, but it holds in general.

38. (Young (1971)). Let $X_{i:n}(\lambda)$ $(1 \le i \le n)$ denote the i'th order statistic in a sample of size n from the gamma distribution given in Exercise 36. Next, let $N_{i:n}(\lambda)$ denote the i'th order statistic, which represents the number of multinomial observations that must be drawn in order to obtain λ observations from each of any i of the n cells. Consider the random variable $X(N_{1:n}(\lambda))$ which is a gamma variable X whose parameter is itself a random variable distributed as the smallest order statistic of the inverse multinomial distribution.

(i) Show that

$$E\left[X^k\{N_{1:n}(\lambda)\}\right] = n^k\, E\{X_{1:n}^k(\lambda)\}$$

and, thence,

$$E\left\{N_{1:n}^{[k]}(\lambda)\right\} = n^k\, E\left\{X_{1:n}^k(\lambda)\right\},$$

where the LHS is the k'th advance factorial moment of $N_{1:n}(\lambda)$.

(ii) Noting now that $N_i(\lambda)$ denoting the number of multinomial observations required to obtain λ observations from the i'th cell are not independent but are exchangeable variates and then by making use of a result due to David and Mishriky (1968), derive the more general relation

$$E\left\{N_{i:n}^{[k]}(\lambda)\right\} = n^k\, E\left\{X_{i:n}^k(\lambda)\right\}, \qquad (i = 1,2,\ldots,n).$$

39. (Arnold (1977)). Let ϕ be a function such that $\phi_{k:n} = E(\phi(X_{k:n}))$ exists for every $k \le n$ for every n. A doubly infinite matrix A with elements a_{ij} will be called an admissible matrix, for a fixed k and n, if

$$g_A(u) = \sum_{i=0}^{\infty} \sum_{j=0}^{\infty} a_{ij}\, u^{i-k+1} (1-u)^{j-n+k} \equiv 1, \qquad u \in [0,1].$$

Verify that if A is admissible for k and n then

$$\phi_{k:n} = \sum_{i=0}^{\infty} \sum_{j=0}^{\infty} a_{ij} \frac{n!\; i!\; j!}{(k-1)!\,(n-k)!\,(i+j+1)!}\, \phi_{i+1:i+j+1}.$$

This result subsumes and extends many of the results of Section 2.1.

40. (Arnold (1977)). For $1 \le k_1 < k_2 < \ldots < k_m \le n$ consider

$$\Psi_{\underline{k}:n} = E(\Psi(X_{k_1:n}, X_{k_2:n}, \ldots, X_{k_m:n}))$$

where Ψ is a function such that $\Psi_{\underline{k}:n}$ exists for all $\underline{k}$ and $\dot{n}$. If

$$g(u) = \sum_{i_1=0}^{\infty} \sum_{i_2=0}^{\infty} \cdots \sum_{i_{m+1}=0}^{\infty} a_{i_1,i_2,\ldots,i_{m+1}} \prod_{j=0}^{m} (u_{j+1}-u_j)^{i_{j+1}-k_{j+1}+k_j+1} \equiv 1$$

where by convention $u_0 = 0$, $u_{m+1} = 1$, $k_0 = 0$ and $k_{m+1} = n+1$, then show that

$$\Psi_{\underline{k}:n} = \sum_{i_1=0}^{\infty} \sum_{i_2=0}^{\infty} \cdots \sum_{i_{m+1}=0}^{\infty} a_{i_1,i_2,\ldots,i_{m+1}}$$

$$\times \frac{n!}{\prod_{j=0}^{m} (k_{j+1}-k_j-1)!} \frac{\prod_{j=0}^{m} i_{j+1}!}{[\, n(\underline{i})\,]!} \Psi_{\underline{k}(\underline{i}):n(\underline{i})}$$

where $\underline{k}(\underline{i}) = (k_1(\underline{i}), k_2(\underline{i}),\ldots,k_m(\underline{i}))$ with $k_j(\underline{i}) = i_1+i_2+\ldots+i_j+j$, $j = 1,2,\ldots,m$, and $n(\underline{i}) = m + \sum_{j=0}^{m} i_{j+1}$. This result subsumes and extends many of the results of Section 2.2.

41. (Arnold (1977)). Show that the results of Exercises 40 and 41 continue to hold when the X_i's are from an exchangeable sequence. Show that they also hold for functions of concomitants of order statistics.

42. (Malik (1967)). For the power–function distribution with probability density function

$$f(x) = ua^{-u} x^{u-1},\ 0 < x \le a,\ a > 0,\ u > 0,$$

show that:

$$\mu_{i:n}^{(k)} = \frac{\Gamma(n+1)}{\Gamma(i)} \frac{\Gamma(i+\frac{k}{u})}{\Gamma(n+1+\frac{k}{u})} a^k, \quad 1 \le i \le n,\ k \ge 1,$$

and

$$\mu_{i,j:n} = \frac{\Gamma(n+1)}{\Gamma(i)} \frac{\Gamma(i+\frac{1}{u})\ \Gamma(j+\frac{2}{u})}{\Gamma(j+\frac{1}{u})\ \Gamma(n+1+\frac{2}{u})} a^2, \quad 1 \le i < j \le n.$$

Show also that the following relations hold in this case:

$$\mu_{i:n}^{(k)} = \frac{i-1+\frac{k}{u}}{i-1} \mu_{i-1:n}^{(k)}, \quad 2 \le i \le n,\ k \ge 1,$$

and

$$\mu_{i,j:n} = \frac{j-1+\frac{2}{u}}{j-1+\frac{1}{u}} \mu_{i,j-1:n}, \quad 1 \le i < j \le n.$$

43. (Hirakawa (1973)). For any arbitrary continuous distribution with median $\tilde{x}$, by defining

$$\Psi_1(k,\ell) = \int_{-\infty}^{\infty} x^k\ [F(x)\{1-F(x)\}]^{\ell}\ f(x)\ dx$$

and

$$\Psi_2(k,\ell) = -\int_{-\infty}^{\tilde{x}} x^k\,[F(x)\{1-F(x)\}]^\ell\,[1-4F(x)\{1-F(x)\}]^{1/2}\,f(x)\,dx$$

$$+\int_{\tilde{x}}^{\infty} x^k\,[F(x)\{1-F(x)\}]^\ell\,[1-4F(x)\{1-F(x)\}]^{1/2}\,f(x)\,dx$$

for $k = 1,2,\ldots,\ \ell = 0,1,2,\ldots$, show that

$$\mu_{n-i+1:n} + \mu_{i:n} = n(n-2i+1)\binom{n-1}{i-1} \sum_{j=0}^{\left[\frac{n-2i+1}{2}\right]} (-1)^j \frac{1}{n-2i+1-j}\binom{n-2i+1-j}{j}\,\psi_1(1,i-1+j)$$

and

$$\mu_{n-i+1:n} - \mu_{i:n} = n\binom{n-1}{i-1} \sum_{j=0}^{\left[\frac{n-2i}{2}\right]} (-1)^j \binom{n-2i-j}{j}\,\Psi_2(1,i-1+j)$$

for $i = 1,2,\ldots,[n/2]$,

and

$$\mu_{i:n} = n\binom{n-1}{i-1}\,\psi_1(1,i-1)$$

for $i = (n+1)/2$ when n is odd.

For continuous distributions symmetric about zero, show also that

$$\mu^{(2k)}_{n-i+1:n} = \mu^{(2k)}_{i:n} = \frac{1}{2}\,n(n-2i+1)\binom{n-1}{i-1} \sum_{j=0}^{\left[\frac{n-2i+1}{2}\right]} (-1)^j \frac{1}{n-2i+1-j}\binom{n-2i+1-j}{j}\,\psi_1(2k,i-1+j)$$

and

$$\mu^{(2k-1)}_{n-i+1:n} = -\,\mu^{(2k-1)}_{i:n} = \frac{1}{2}\,n\binom{n-1}{i-1} \sum_{j=0}^{\left[\frac{n-2i}{2}\right]} (-1)^j \binom{n-2i-j}{j}\,\psi_2(2k-1,i-1+j)$$

for $i = 1,2,\ldots,[n/2],\ k = 1,2,\ldots,$

and

$$\mu^{(2k)}_{i:n} = n\binom{n-1}{i-1}\,\psi_1(2k,i-1), \qquad \mu^{(2k-1)}_{i:n} = 0$$

for $i = (n+1)/2$ when n is odd.

44. (Balakrishnan, 1989a). By making use of Relations 2.2 and 2.13, show that for any arbitrary distribution

$$(i-1)\,\sigma_{i,j:n} + (j-i)\,\sigma_{i-1,j:n} + (n-j+1)\,\sigma_{i-1,j-1:n}$$
$$= n\,\sigma_{i-1,j-1:n-1} + n(\mu_{i-1:n-1} - \mu_{i-1:n})\,(\mu_{j-1:n-1} - \mu_{j:n})$$

for $2 \le i < j \le n$.

45. (Stigler (1977); Rohatgi and Saleh (1988)). Let F be an arbitrary distribution function. Then for any real number α the Newton's binomial series expansion gives

$$1 = \{F + (1-F)\}^\alpha = \sum_{r=0}^{\infty} \binom{\alpha}{r} F^r\,(1-F)^{\alpha-r}$$

where

$$\begin{bmatrix}\alpha\\ r\end{bmatrix} = \frac{\alpha(\alpha-1)\ \cdots\ (\alpha-r+1)}{r!}, \qquad \begin{bmatrix}\alpha\\ 0\end{bmatrix} = 1.$$

Then

$$F_{i:\alpha} = \sum_{r=i}^{\infty} \begin{bmatrix}\alpha\\ r\end{bmatrix} F^r (1-F)^{\alpha-r}$$

defines a distribution function. When $\alpha = n$ is an integer, it is just the cdf of the i'th order statistic in a sample of size n from F. Hence, $F_{i:\alpha}$ may be considered as the cdf of the i'th order statistic $X_{i:\alpha}$ with nonintegral sample size α. Then most of the results given in this Chapter may be generalized to this case.

In particular, prove that the triangle rule

$$i\, F_{i+1:\alpha} + (\alpha-i)\, F_{i:\alpha} = \alpha\, F_{i:\alpha-1}$$

holds for $i \le \alpha-1$.

46. (Balakrishnan, 1989c). (i) By extending the argument that has been used to prove the relation in (2.68), show that for the order statistics from n exchangeable variables

$$\begin{bmatrix}n\\ m\end{bmatrix} \mu^{(k)}_{i:m} = \sum_{r=0}^{n-m} \begin{bmatrix}i+r-1\\ r\end{bmatrix} \begin{bmatrix}n-i-r\\ m-i\end{bmatrix} \mu^{(k)}_{i+r:n}$$

for $1 \le i \le m \le n-1$. We may note here that this generalizes the relation due to Sillitto (1964) (also presented in Exercise 2) to the case when the order statistics arise from n exchangeable random variables.

(ii) Similarly, show that for $1 \le i < j \le m \le n-1$

$$\begin{bmatrix}n\\ m\end{bmatrix} \mu_{i,j:m} = \sum_{r=0}^{n-m} \sum_{s=r}^{n-m} \begin{bmatrix}i-1+r\\ r\end{bmatrix} \begin{bmatrix}j-i-1+s-r\\ s-r\end{bmatrix} \begin{bmatrix}n-j-s\\ n-m-s\end{bmatrix} \mu_{i+r,j+s:n}.$$

CHAPTER 3

BOUNDS ON EXPECTATIONS OF ORDER STATISTICS

3.0. Introduction

The classic results on universal bounds for order statistics were provided in papers published simultaneously by Gumbel (1954) and Hartley and David (1954). Antecedents and partial anticipations can be identified, particularly noteworthy is the contribution of Plackett (1947). These authors all dealt with the i.i.d. case. Relaxation of the identical distribution and the independence assumptions was not explicitly treated until 25 years later, though again one can identify relevant insights throughout the intervening period. Two papers which turned out to be influential in refocussing attention on variations on the Gumbel–Hartley–David theme were Samuelson (1968) and Lai and Robbins (1976). Samuelson's note, with its irresistable title "How deviant can you be" spawned a torrent of generalizations, several of which referred to bounds on order statistics. It also spawned a flurry of rediscoveries of earlier notes on these topics. Ultimate priority seems hard to pin down although Scott's (1936) appendix to the Pearson and Chandra Sekar paper stands out as one of the earliest sources thus far identified. Lai and Robbins (1976) introduced a class of maximally dependent joint distributions. The name maximally dependent is perhaps an infelicitous choice but apparently we are stuck with it. In any case, such joint distributions conveniently provide extreme cases for distributions of possibly dependent maxima. Sections 1 through 3 will survey the universal bounds obtainable using all the aforementioned techniques.

Section 4 will focus on bounds on expectations of order statistics when the parent distributions are assumed to belong to specific restricted families (unimodal, IFR or perhaps more specifically, normal).

An alternate title for this chapter will suggest itself as the story unfolds. It might well have been entitled: "1001 ways to use the Schwarz inequality".

3.1. Universal bounds in the i.i.d. case

Suppose $X_1, X_2, \ldots, X_n$ are i.i.d. random variables with common distribution function F. We assume that F has finite mean μ and finite variance σ^2. Subject only to this moment restriction, we seek bounds on expectations of functions of the order statistics $X_{1:n}, \ldots, X_{n:n}$.

The technique is well illustrated by derivation of a bound for $E(X_{n:n})$ $(=\mu_{n:n})$. Without loss of generality translate or rescale the X_i's so that $E(X) = 0$ and $E(X^2) = 1$. We may write

$$E(X_{n:n}) = \int_0^1 F^{-1}(u) n u^{n-1} du \tag{3.1}$$

where F^{-1} is as defined in equation (1.3). The Schwarz inequality for square integrable functions g and h on [0,1] takes the form

$$\int_0^1 g(u)h(u)du \le \sqrt{\int_0^1 g^2(u)du \int_0^1 h^2(u)du} \tag{3.2}$$

with equality if and only if $g = kh$ a.e. on the set where $gh > 0$.

Before applying the Schwarz inequality we rewrite (3.1) in the form

$$E(X_{n:n}) = \int_0^1 F^{-1}(u)[nu^{n-1} - c]du. \tag{3.3}$$

Expression (3.3) is valid for every real c, since $\int_0^1 F^{-1}(u)du = 0$. To apply the Schwarz inequality in (3.3), we identify

$$g(u) = F^{-1}(u)$$

and

$$h(u) = nu^{n-1} - c.$$

Since $E(X^2) = \int_0^1 [F^{-1}(u)]^2 du = 1$, this yields

$$E(X_{n:n}) \le \sqrt{\int_0^1 (nu^{n-1} - c)^2 du}$$

$$= \sqrt{c^2 - 2c + n^2(2n-1)^{-1}}.$$

This bound is smallest when $c = 1$. Setting $c = 1$ we have

$$E(X_{n:n}) \le (n-1)(2n-1)^{-1/2}. \tag{3.4}$$

Equality will obtain in (3.4) if the common distribution function of the X_i's has an inverse which satisfies

$$F^{-1}(u) = k[nu^{n-1} - 1], \ 0 < u < 1. \tag{3.5}$$

The constant k in (3.5) must be chosen such that $E(X^2) = \int_0^1 [F^{-1}(u)]^2du = 1$. It follows that

$$k = \sqrt{2n-1} / (n-1). \tag{3.6}$$

From (3.5) with k given by (3.6) we find the extremal distribution is of the form

$$F(x) = [(1 + \tfrac{x}{k})/n]^{\frac{1}{n-1}}, \ -k < x < \sqrt{2n-1}, \tag{3.7}$$

where k is as in (3.6). Observe that when $n = 2$, (3.7) reduces to a uniform distribution on the interval $(-\sqrt{3}, \sqrt{3})$. Graphs of the distribution (3.7) and the corresponding densities are provided by Gumbel (1954) for the cases $n = 2,3,4,5$.

The extremal distributions given by (3.7) (for $n > 2$) are rather unusual. In many situations, additional information about F might suggest that an extremal value for $E(X_{n:n})$ might be considerably less than the value provided by (3.4). For example we might know that the common distribution of the X_i's is symmetric. This problem was actually treated by Moriguti (1951) before the general case was resolved.

The requirement that F be symmetric can be written in terms of the inverse distribution function as follows

$$F^{-1}(u) = -F^{-1}(1-u), \quad 0 \le u \le 1. \tag{3.8}$$

If F is to have mean 0 and variance 1 then, in addition to (3.8), the only additional requirement is that

$$\int_{1/2}^{1} [F^{-1}(u)]^2\, du = \frac{1}{2}. \tag{3.9}$$

To determine the maximal value of $E(X_{n:n})$ for such a symmetric parent distribution we need to maximize (3.1) subject to (3.8) and (3.9). Equation (3.1) may be rewritten, using (3.8), as

$$E(X_{n:n}) = \int_{1/2}^{1} F^{-1}(u)\, n[u^{n-1} - (1-u)^{n-1}]du. \tag{3.10}$$

The Schwarz inequality (3.2) may be applied to (3.10) using the choices

$$g(u) = F^{-1}(u)\, I(u > \tfrac{1}{2})$$

and

$$h(u) = n[u^{n-1} - (1-u)^{n-1}]\, I(u > \tfrac{1}{2}).$$

Using (3.9) this yields

$$\begin{aligned} E(X_{n:n}) &\le \sqrt{\tfrac{1}{2} \textstyle\int_{1/2}^{1} n^2[u^{n-1} - (1-u)^{n-1}]^2\, du} \\ &= \frac{n}{\sqrt{2}} \sqrt{\frac{1}{2n-1} - B(n,n)} \end{aligned} \tag{3.11}$$

where $B(\cdot,\cdot)$ is the classical Beta function, (Moriguti (1951)). The bound (3.11) is achievable by a distribution whose inverse is proportional to $n[u^{n-1} - (1-u)^{n-1}]$ on the interval (1/2, 1) and is extended to (0, 1/2) using (3.8). The required constant of proportionality is determined by the requirement that var(X) = 1 (i.e. (3.9) must hold). Moriguti supplied graphs of the corresponding extremal densities for n = 2,3,4,5,-6,8,10. It is interesting to observe that in both of the cases n = 2 and 3, the extremal distribution is uniform $(-\sqrt{3}, \sqrt{3})$.

The extremal inverse distribution function (for which the bound (3.11) is achieved) is of the form

$$F^{-1}(u) = k[u^{n-1} - (1-u)^{n-1}], \quad 0 \le u \le 1 \tag{3.12}$$

where

$$k = \frac{1}{\sqrt{2}} \left[\frac{1}{2n-1} - B(n,n)\right]^{-1/2}. \tag{3.13}$$

From (3.12), it is clear that the support of the extremal distribution function is (−k, k) where k is given by (3.13). A closed form for the extremal distribution is not usually available (the exception being the cases n = 2 and 3, alluded to above).

For an arbitrary parent distribution function F the expected range of a sample of size n is given by

$$E(X_{n:n} - X_{1:n}) = \int_0^1 F^{-1}(u)\, n[u^{n-1} - (1-u)^{n-1}]\, du. \tag{3.14}$$

Evidently the Schwarz inequality can be applied and, subject to the requirement that $\int_0^1 [F^{-1}(u)]^2 du = 1$, inverse distributions of the form (3.12) will maximize the expected range (among the class of all possible parent distributions, symmetric or not). The bound obtained in this way is

$$E(X_{n:n} - X_{1:n}) \le n\sqrt{2} \sqrt{\frac{1}{2n-1} - B(n,n)}, \tag{3.15}$$

exactly two times the bound (3.11) (Plackett (1947)).

In Moriguti (1951) rather complicated lower bounds are presented for $var(X_{n:n})$ and $var(X_{n:n})/[E(X_{n:n})]^2$ assuming the parent distribution is symmetric about zero. Again the Schwarz inequality is the key tool. In a later paper, Moriguti (1953), he considers bounds on the expectation of the i'th order statistic. We may write

$$E(X_{i:n}) = \int_0^1 F^{-1}(u)\, g_{i:n}(u)\, du \tag{3.16}$$

where $g_{i:n}(u)$ is the density of the i'th order statistic from a uniform (0,1) distribution. Mimicking the argument which led from (3.1) to the bound (3.4), we find subject to $E(X) = 0$ and $E(X^2) = 1$ that

$$|E(X_{i:n})| \le \left[n \frac{\binom{2n-2i}{n-i}\binom{2i-2}{i-1}}{\binom{2n-1}{n-1}} - 1 \right]^{1/2} . \tag{3.17}$$

Equality in (3.17) would occur if $F^{-1}(u) \propto g_{i:n}(u) - 1$. Since $g_{i:n}(u)$ is only monotone for $i = 1$ or n, the bound (3.17) is only sharp in these cases (F^{-1} being itself monotone cannot be proportional to a non-monotone function). Moriguti suggests an ingenious way to determine a sharp bound. Simply replace $g_{i:n}$ by $h_{i:n}$ which is an increasing density chosen in the following fashion. Consider all distributions on [0,1] corresponding to random variables which are stochastically larger than $U_{i:n}$ and possess increasing densities. Let $H_{i:n}$ be the supremum of this class of distributions and let $h_{i:n}$ be the corresponding density (in Moriguti's terms $H_{i:n}$ is the greatest convex minorant of $G_{i:n}(= F_{U_{i:n}})$).

Rather than consider a single order statistic we might wish to bound expectations of certain linear combinations of order statistics. The Schwarz inequality technique can sometimes be used here. Nagaraja (1981) obtained bounds on the expected selection differential, i.e. $E\left[\frac{1}{k} \sum_{j=n-k+1}^{n} X_{j:n}\right]$ (see Exercise 33).

Sugiura (1962) observed that the bound (3.17) can be thought of as the first term in an expansion for $E(X_{i:n})$ based on orthogonal polynomials. The argument may be presented with reference to <u>any</u> orthonormal system over (0,1) but there are certain advantages associated with the selection of Legendre polynomials. A sequence of functions $\{\phi_k(u)\}_{k=0}^{\infty}$ is a complete orthonormal system in $L^2(0,1)$ (the class of square integrable functions on (0,1)) if

$$\int_0^1 \phi_j(u)du = 0, \quad \int_0^1 \phi_j^2(u)du = 1,$$

$$\int_0^1 \phi_j(u)\, \phi_\ell(u)du = 0, \quad j \ne \ell$$

and for every $f \in L^2(0,1)$ we have

$$f = \lim_{n\to\infty} \sum_{j=0}^{n} a_j \phi_j \tag{3.18}$$

and

$$\int_0^1 f^2(u)du = \sum_{j=0}^{\infty} a_j^2 \tag{3.19}$$

where

$$a_j = \int_0^1 f(u)\phi_j(u)du. \tag{3.20}$$

If we take two members of $L^2(0,1)$, say $f = \sum_{j=0}^{\infty} a_j\phi_j$ and $g = \sum_{j=0}^{\infty} b_j\phi_j$ [where $b_j = \int g\phi_j$] then for any k

$$\left|\int_0^1 f(u)g(u)du - \sum_{j=0}^{k} a_j b_j\right|$$

$$= \left|\sum_{j=k+1}^{\infty} a_j b_j\right| \le \sqrt{\left[\sum_{j=k+1}^{\infty} a_j^2\right]\left[\sum_{j=k+1}^{\infty} b_j^2\right]} \tag{3.21}$$

$$= \sqrt{\left[\int_0^1 f^2(u)du - \sum_{j=0}^{k} a_j^2\right]\left[\int_0^1 g^2(u)du - \sum_{j=0}^{k} b_j^2\right]}.$$

Relation (3.18) and the Schwarz inequality were used in this augument. Equality will obtain in (3.21) if $a_j = cb_j$, $\forall\, j > k$, i.e. if

$$f - \sum_{j=0}^{k} a_j\phi_j \propto g - \sum_{j=0}^{k} b_j\phi_j. \tag{3.22}$$

Although in (3.21) the first j coefficients were used this was not important in the argument. As Joshi (1969) points out, we have for any subset J of the set {0,1,2,...},

$$\left|\int_0^1 f(u)g(u)du - \sum_{j\varepsilon J} a_j b_j\right|$$

$$\le \sqrt{\left[\int_0^1 f^2(u)du - \sum_{j\varepsilon J} a_j^2\right]\left[\int_0^1 g^2(u)du - \sum_{j\varepsilon J} b_j^2\right]}. \tag{3.23}$$

The Legendre polynomials on (0,1) are defined by

$$\phi_j(u) = \frac{\sqrt{2j+1}}{j!}\frac{d^j}{du^j}\, u^j(u-1)^j, \quad j = 0,1,2,.... \tag{3.24}$$

The first three are specifically

$\phi_0(u) = 1$

$\phi_1(u) = \sqrt{3}\,(2u-1)$

$\phi_2(u) = \sqrt{5}\,(6u^2 - 6u + 1).$

The Sugiura bounds on $E(X_{i:n})$ are included in the following.

Theorem 3.1: Let $X_{i:n}$ denote the i'th order statistic of a sample from the distribution F with mean μ and variance σ^2. Let $\{\phi_0,...,\phi_k\}$ be an orthonormal system in $L^2(0,1)$ with $\phi_0 \equiv 1$ then

$$\left|E(X_{i:n}) - \mu - \sum_{j=1}^{k} a_j b_j\right|$$

$$\le \sqrt{\left[\sigma^2 - \sum_{j=1}^{k} a_j^2\right]\left[\frac{B(2i-1,2n-2i+1)}{[B(i,n-i+1)]^2} - 1 - \sum_{j=1}^{k} b_j^2\right]} \tag{3.25}$$

where

$$a_j = \int_0^1 F^{-1}(u)\phi_j(u)du$$

and

$$b_j = [B(i,n-i+1)]^{-1} \int_0^1 u^{i-1}(1-u)^{n-i}\phi_j(u)du.$$

Proof: Apply (3.21) with $f(u) = F^{-1}(u)$ and $g(u) = [B(i,n-i+1)]^{-1} u^{i-1}(1-u)^{n-i}$. With these definitions, $E(X_{i:n}) = \int_0^1 f(u)g(u)du$ and the result follows since $a_0 = \int_0^1 F^{-1}(u)\phi_0(u)du = \mu$ and $b_0 = \int_0^1 g(u)\phi_0(u)du = 1$ (recall $\phi_0(u) \equiv 1$).

If the parent distribution is assumed to be symmetric then $a_j = 0 \ \forall\, j$ even and applying (3.23) with $J = \{1,3,\ldots,2k+1,0,2,4,6,\ldots\}$ we find

$$\left|E(X_{i:n}) - \sum_{j=0}^{k} a_{2j+1}\, b_{2j+1}\right| \le \sqrt{\left[\sigma^2 - \sum_{j=0}^{k} a_{2j+1}^2\right]\left[\frac{B(2i-1,2n-2i+1)}{[B(i,n-i+1)]^2} - \sum_{j\varepsilon J} b_j^2\right]}. \tag{3.26}$$

However

$$\sum_{j \text{ even}} b_j^2 = \int_0^1 [g(u) + g(1-u)]^2/4\, du = \frac{B(2i-1,2n-2i+1) + B(n,n)}{2[B(i,n-i+1)]^2}$$

yielding the result

$$\left|E(X_{i:n}) - \sum_{j=0}^{k} a_{2j+1}\, b_{2j+1}\right| \le \sqrt{\left[\sigma^2 - \sum_{j=0}^{k} a_{2j+1}^2\right]\left[\frac{B(2i-1,2n-2i+1) - B(n,n)}{2[B(i,n-i+1)]^2} - \sum_{j=0}^{k} b_{2j+1}^2\right]}. \tag{3.27}$$

In general the bounds (3.25) and (3.27) cannot be expected to be sharp.

Joshi (1969) discusses analogs of (3.25) and (3.27) starting with $f(u) = F^{-1}(u)u^p(1-u)^q$ rather than $f(u) = F^{-1}(u)$. This program yields bounds on $E(X_{i:n})$ in terms of the second moments of another order statistic. His bounds may thus be used in some cases when var(X) does not exist, e.g. samples from a Cauchy distribution. This program is discussed in more detail in Chapter 4.

3.2. Variations on the Samuelson–Scott theme

Samuelson's (1968) remark that no unit in a finite population of N elements can lie more than $\sqrt{N-1}$ standard deviations away from the population mean refocussed attention on a result implicitly known and occassionally remarked upon in the literature for decades. Perhaps the earliest explicit statement and proof of the result is that supplied by Scott (1936) in an appendix to a paper on outliers by Pearson and Chandra Sekar (1936). Let us begin by reviewing Scott's result.

Focus on a finite population of N units each possesses a numerical attribute x_i, $i = 1,2,\ldots,N$. Denote by $x_{i:N}$ the i'th largest of the x_i's (i.e. $x_{1:n}$ is the smallest, etc.). Denote the population mean and the population variance by $\bar{x}$ and s^2. Thus

$$\bar{x} = \frac{1}{N} \sum_{i=1}^{N} x_i \tag{3.28}$$

and

$$s^2 = \frac{1}{N} \sum_{i=1}^{N} (x_i - \bar{x})^2. \tag{3.29}$$

It is convenient to introduce notation for deviations, absolute deviations and ordered absolute deviations. Thus

$$d_i = x_i - \bar{x}, \quad i = 1,2,\ldots,N, \tag{3.30}$$

$$\tau_i = \left| x_i - \bar{x} \right|, \quad i = 1,2,\ldots,N \tag{3.31}$$

and

$$0 \le \tau_{1:N} \le \tau_{2:N} \le \cdots \le \tau_{N:N}. \tag{3.32}$$

Scott provides bounds for $\tau_{i:n}$ $(i = 1,2,\ldots,N)$.

Theorem 3.2 (Scott, 1936): In a finite population of N elements if $\bar{x}$, s^2 and $\tau_{i:n}$ are as defined above, we have

$$(i \neq 1,\ N-i+1 \text{ odd})\ \tau_{i:N} \le s \sqrt{\frac{(i-1)N}{(N-i+1)(i-1) + 1}} \tag{3.33}$$

$$(i = 1,\ N \text{ odd})\ \tau_{1:N} \le s \sqrt{\frac{N-1}{N(N+1)}} \tag{3.34}$$

$$(N-i+1 \text{ even})\ \tau_{i:N} \le s \sqrt{\frac{1}{N-i+1}}. \tag{3.35}$$

(Remark: Samuelson's inequality corresponds to the case $i = N$ in (3.33).)

Proof: Scott's ingenious constructive proof is apparently the only proof available in the literature. Rather than reproduce it modulo notation changes we will merely list examples of extremal populations in which equality obtains in (3.33) – (3.35), inviting the reader to try a hand at developing an alternative proof.

Case (i) $(i \neq 1,\ N-i+1$ odd).

Let $a_{N,i} = \sqrt{\frac{N(i-1)}{(i-1)(N-i+1) + 1}}$.

Take $\frac{1}{2}(N-i)$ of the x's to be $a_{N,i}$, take $\frac{1}{2}(N-i) + 1$ of the x's to be $-a_{N,i}$ and take the remaining x's to be equal to $c_{N,i}$ such that $\sum_{i=1}^{N} x_i = 0$. Equality then obtains in (3.33).

Case (ii) $(i = 1,\ N$ odd).

Let $a_{N,1} = \sqrt{(N+1)/(N-1)}$.

Take $\frac{1}{2}(N-1)$ of the x's to be $a_{N,1}$ and take the other x's to be equal to $-a_{N,1}^{-1}$. Equality then obtains in (3.34).

Case (iii) $(N-i+1$ even).

Let $b_{N,i} = \sqrt{\frac{1}{N-i+1}}$.

Take $\frac{1}{2}$ (N–i+1) of the x's equal to $b_{N,i}$, take $\frac{1}{2}$ (N–i+1) of the x's equal to $-b_{N,i}$ and take the remaining x's to be zero. Equality then obtains in (3.35).

A deeper understanding of Scott's proof may be obtained by perusing the generalization derived by Beesack (1976) (see Exercises 4–6). Scott's results regarding ordered absolute deviations are naturally of interest in the outlier detection scenario. They are important in determining the range of certain natural outlier detecting test statistics. Typically the cases $i = N$, $N - 1$, and $N - 2$ are of interest (corresponding to one, two or three possible outliers).

It is instructive to focus on the case $i = N$ (the Samuelson case) and consider several alternative proofs. The alternative proofs often suggest different possible extensions of the result. The Schwarz inequality may be perceived to be lurking in the background of many of the proofs.

Theorem 3.3: Let $x_1, x_2, \ldots, x_N$ be N real numbers then

$$\max |x_i - \bar{x}| \le s\sqrt{N-1} \tag{3.36}$$

where $\bar{x}$ and s are defined in equations (3.28) and (3.29).

First proof: (Basically Samuelson (1968) and Scott (1936)). Fix x_1. Replacing the other n–1 observations by their mean will reduce s and leave $\bar{x}$ unchanged. Thus x_1 will be a, say, and the other x's will be equal to b. For such a configuration it is easily verified that (3.36) holds.

Second proof: (Arnold (1974), Dwass (1975)). Fix i. We have

$$\begin{aligned}(x_i - \bar{x})^2 &= \left[\sum_{j\neq i} (x_i - \bar{x})\right]^2 \\ &\le \left[\sum_{j\neq i} (x_i - \bar{x})^2\right] (N-1) \\ &= N(N-1)s^2 - (N-1)(x_i - \bar{x})^2.\end{aligned}$$

Thus $|x_i - \bar{x}| \le s\sqrt{N-1}$ and this clearly holds for every i, so (3.36) is verified.

Third proof: (Kempthorne (1973), see also Nair (1948)). Let O be a Helmert orthogonal $N \times N$ matrix with first and last rows given by

$$\left[\frac{1}{\sqrt{N}}, \frac{1}{\sqrt{N}}, \ldots, \frac{1}{\sqrt{N}}\right]$$

and

$$\left[\frac{1}{\sqrt{N(N-1)}}, \ldots, \frac{1}{\sqrt{N(N-1)}}, \frac{-(N-1)}{\sqrt{N(N-1)}}\right]. \tag{3.37}$$

Let $\underline{x} = (x_1, \ldots, x_N)$ and define $\underline{y}$ by $\underline{y} = 0\underline{x}$. Since 0 is orthogonal we have

$$\begin{aligned}\sum_{i=1}^{N} x_i^2 &= \sum_{i=1}^{N} y_i^2 \ge y_1^2 + y_N^2 \\ &= N\bar{x}^2 + \frac{[(N\bar{x} - x_N) - (N-1)x_N]^2}{N(N-1)}.\end{aligned}$$

It follows that

$$\frac{N(x_N - \bar{x})^2}{(N-1)} \le \sum_{i=1}^{N} x_i^2 - N\bar{x}^2 = Ns^2.$$

Thus $|x_N - \bar{x}| \le s\sqrt{N-1}$. Analogously $|x_i - \bar{x}| \le s\sqrt{N-1}$ $\forall$ i and (3.36) follows.

Fourth proof: (Arnold (1974)). Assume without loss of generality that $\bar{x} = 0$. Consider the sequence of N random variables obeying the regression model

$$Y_i = \alpha + \beta x_i + \varepsilon_i, \quad i = 1,2,\ldots,N \tag{3.38}$$

where the ε_i's are i.i.d. random variables with common finite variance σ^2. The least squares estimates of α and β are respectively

$$\hat{\alpha} = \frac{1}{N} \sum_{i=1}^{N} Y_i$$

and

$$\hat{\beta} = \left[\sum_{i=1}^{N} x_i Y_i\right] \Big/ \left[\sum_{i=1}^{N} x_i^2\right]. \tag{3.39}$$

The i'th residual say Z_i is defined by

$$Z_i = Y_i - \hat{\alpha} - \hat{\beta} x_i. \tag{3.40}$$

It is readily verified that

$$\operatorname{var}(Z_i) = \sigma^2 \left\{1 - \frac{1}{N} - \frac{x_i^2}{\sum_{i=1}^{N} x_i^2}\right\}. \tag{3.41}$$

Variances are non–negative. The non–negativity of (3.41) is equivalent to (3.36).

Fifth proof: (O'Reilly (1975, 1976)). Let $y_1,\ldots,y_N$ be row vectors in $\mathbb{R}^2$ and let Y be the $N \times 2$ matrix whose rows are the y_i's. O'Reilly (1975) shows that for any $y = \sum_{i=1}^{N} \alpha_i y_i$ with $\sum_{i=1}^{N} \alpha_i^2 = 1$ we have

$$y'(Y'Y)^{-1}\, y \le 1. \tag{3.42}$$

Consequently for any i,

$$y_i(Y'Y)^{-1}\, y_i \le 1. \tag{3.43}$$

Let $y_i = (1,x_i)$, $i = 1,2,\ldots,N$ then (3.36) follows from (3.43).

Sixth proof: (Smith (1980)). Without loss of generality $\sum_{i=1}^{N} x_i = 0$ and $\frac{1}{N}\sum_{i=1}^{N} x_i^2 = 1 = s^2$. Assume the x_i's are arranged in increasing order so that x_N is the largest. Consider a random variable X defined by $P(X = x_i) = \frac{1}{N}$, $i = 1,2,\ldots,N$. A version of the Cantelli inequality is that, since $E(X) = 0$ and $\operatorname{var}(X) = 1$, we have for any $x > 0$,

$$P(X \ge x) \le (1 + x^2)^{-1}. \tag{3.44}$$

Let $x = x_N$ in (3.44) and we have

$$\frac{1}{N} \le (1 + x_N^2)^{-1}. \tag{3.45}$$

From this it follows that $x_N^2 \le (N-1) = (N-1)s^2$. Analogously (by considering $y_i = -x_i$) we have $x_1^2 \le (N-1)s^2$ and (3.36) then follows.

Other proofs can undoubtedly be unearthed but the above list is representative. (See Exercise 29 where the less stringent bound $s\sqrt{N}$ is discussed).

A p–dimensional version of Theorem 3.3 is clearly possible. The regression argument (the fourth proof) extends easily.

Theorem 3.4: Suppose each element in a finite population with N elements has p measurable attributes. For element i, denote these attributes by $(x_{i1},\ldots,x_{ip})$ then

$$\max_i d_i' \Sigma^{-1} d_i \le N - 1 \tag{3.46}$$

where Σ is the population variance covariance matrix (assumed non–singular) and d_i is the vector of deviations of the attributes of element i from the corresponding population means (i.e. $d_{ik} = x_{ik} - \bar{x}_{.k}$).

Rather than present a proof of (3.46) we will consider a more general problem (following Arnold and Groeneveld (1974)). Consider a general linear model $Y = Z\beta + \varepsilon$ where Z is an $N \times (p+1)$ full rank matrix, $\beta \varepsilon \mathbb{R}^{p+1}$ and ε is a vector of i.i.d. random variables each with mean zero and variance 1. The least squares estimates of β are

$$\hat{\beta} = (Z'Z)^{-1} Z'Y \tag{3.47}$$

and the corresponding vector of residuals is

$$\hat{e} = Y - Z\hat{\beta}$$

$$= [I - Z(Z'Z)^{-1} Z']Y. \tag{3.48}$$

Since the variance covariance matrix of $\hat{e}$ is non–negative definite, it follows that for any $\underline{\lambda} \varepsilon \mathbb{R}^N$,

$$\underline{\lambda}'\{I - Z(Z'Z)^{-1} Z'\} \underline{\lambda} \ge 0 \tag{3.49}$$

(this, incidently, gives an alternative proof of (3.43)).

Now consider our population of N units with p measurable attributes. Define

$$X = (x_{ij})_{N\times p} \tag{3.50}$$

and

$$D = (d_{ij})_{N\times p} \tag{3.51}$$

where $d_{ij} = x_{ij} - \bar{x}_{.j}$. We may denote the population variance covariance matrix by

$$\Sigma = \frac{1}{N} (D'D). \tag{3.52}$$

We have:

Theorem 3.5: (Arnold and Groeneveld (1974)). For any $\underline{\lambda} \varepsilon \mathbb{R}^N$,

$$(\underline{\lambda}'D)\Sigma^{-1}(\underline{\lambda}'D)' \le N \sum_{i=1}^{N} \lambda_i^2 - \left[\sum_{i=1}^{N} \lambda_i\right]^2$$

$$= N \sum_{i=1}^{N} (\lambda_i - \bar{\lambda})^2 \tag{3.53}$$

Proof: Take $Z = (1,D)$ in (3.49).

Remark: Results similar in spirit to Theorem 3.5 but in different contexts may be found in Prescott (1977), Loynes (1979) and Arnold and Groeneveld (1978).

For any two vector $\underline{\alpha}$ and $\underline{\beta}$ in $\mathbb{R}^p$ we define the squared generalized distance between $\underline{\alpha}$ and $\underline{\beta}$ to be

$$d_\Sigma^2(\underline{\alpha},\underline{\beta}) = (\underline{\alpha} - \underline{\beta})\Sigma^{-1}(\underline{\alpha} - \underline{\beta})' \tag{3.54}$$

Using this definition we may state the following immediate corollaries of Theorem 3.5.

Corollary 3.6: In a finite population with N elements each with p attributes, the squared generalized distance between the vector of attribute means based on a sample of size n and the vector of population means for the attributes cannot exceed $(N/n) - 1$.

Proof: The mean vector for a sample of size n is expressible as $\underline{\lambda}'X$ where $\underline{\lambda}$ has n of its coordinates equal to $(1/n)$ and the remaining ones are equal to 0. Then $\underline{\lambda}'X - \bar{x}^{(N)} = \underline{\lambda}'D$ (where $\bar{x}^{(N)}$ is the vector of population means of the attributes) and the result follows from (3.53). Theorem 3.4 corresponds to the case $n = 1$. Theorem 3.3 corresponds to the case $n = 1$ and $p = 1$.

Corollary 3.7: Consider a finite population of N elements each with p attributes. The squared generalized distance between the vectors of attribute means based on two possible overlapping samples of sizes n_1 and n_2 respectively, cannot exceed $N(n_1^{-1} + n_2^{-1})$.

Proof: Let $\bar{x}^{(n_1)}$ and $\bar{x}^{(n_2)}$ be the vectors of sample means of the attributes and let ℓ be the number of elements common to the samples. The squared generalized distance between $\bar{x}^{(n_1)}$ and $\bar{x}^{(n_2)}$ is $(\underline{\lambda}'X)\Sigma^{-1}(\underline{\lambda}'X)'$ where $\underline{\lambda}$ has ℓ entries equal to $n_1^{-1} - n_2^{-1}$, $n_1 - \ell$ entries equal to n_1^{-1}, $n_2 - \ell$ entries equal to n_2^{-1} and the remaining entries equal to 0. Since $\sum_{i=1}^{N} \lambda_i = 0$ we have $\underline{\lambda}'X = \underline{\lambda}'D$ and the result follows from (3.58).

Observe that Corollary 3.7, in the case $n_1 = n_2 = 1$, gives an upper bound, 2N, on the squared generalized distance between the attribute vectors of any two units in the population.

Theorem 3.5 has several interesting interpretations in the one dimensional case ($p = 1$). Suppose $\underline{\lambda}$ has 2 non–zero entries, a 1 in coordinate i and a –1 in coordinate j. In this situation (3.58) yields

$$\left|x_i - x_j\right| \le s\sqrt{2N} \qquad (3.55)$$

where s^2 is as defined in (3.29). Since x_i could be the smallest of the x's, i.e. $x_{1:N}$, and x_j the largest, $x_{N:N}$, this yields the Nair (1948) – Thomson (1955) bound on the range of the population

$$R = x_{N:N} - x_{1:N} \le s\sqrt{2N}. \qquad (3.56)$$

In the one–dimensional case, let us rewrite (3.58) in terms of the ordered x_i's. We thus have, for any $\underline{\lambda} \in \mathbb{R}^N$,

$$\left|\sum_{i=1}^{N} \lambda_i(x_{i:N} - \bar{x})\right| \le s\sqrt{N \sum_{i=1}^{N} (\lambda_i - \bar{\lambda})^2} \qquad (3.57)$$

(a result presumably first derived by Nair (1948) in the case $\sum\lambda_i = 0$). This result yields the following set of bounds on the $x_{i:N}$'s.

Theorem 3.8: For $i = 1,2,\ldots,N$

$$-s\sqrt{\frac{N-i}{i}} \le x_{i:N} - \bar{x} \le s\sqrt{\frac{i-1}{N-i+1}}. \qquad (3.58)$$

Proof: Obviously

$$x_{i:N} - \bar{x} \le \frac{1}{N-i+1} \sum_{j=i}^{N} (x_{j:N} - \bar{x}).$$

If we apply (3.57) with $\underline{\lambda}$ such that $\lambda_1 = \ldots = \lambda_{i-1} = 0$, $\lambda_i = \lambda_{i+1} = \ldots = \lambda_N = (N - i + 1)^{-1}$, we conclude that

$$x_{i:N} - \bar{x} \leq s\sqrt{\frac{i - 1}{N - i + 1}}. \tag{3.59}$$

If we define $y_i = - x_i$ then $s_y^2 = s^2$ and using (3.59) on the y_i's with $i = N-i+1$ we have

$$\bar{x} - x_{i:N} = \bar{y}_{N-i+1:N} - y \leq s\sqrt{\frac{(N-i+1) - 1}{N - (N-i+1) + 1}}$$
$$= s\sqrt{\frac{N - i}{i}}$$

thus confirming the left hand inequality in (3.58).

Theorem 3.8 is implicit in Arnold and Groeneveld (1979), explicit in Wolkowicz and Styan (1979), implicit in Mallows and Richter (1969) and explicit in Hawkins (1971) and Boyd (1971). Scott (1936), without proof, gives the result $x_{N-1:N} - \bar{x} \leq \sqrt{(N - 2)/2}$, i.e. the upper bound in (3.58) in the case $i = N - 1$.

Most of the bounds given in Theorem 3.8 are best possible in the sense that there exist populations in which the bounds are achieved (see Exercise 9). The exceptions are the two trivial bounds, the upper bound for $x_{1:N} - \bar{x}$ (namely 0) and the lower bound for $x_{N:N} - \bar{x}$ (again 0). Both of these can be improved.

Theorem 3.9: (Hawkins (1971))

$$x_{1:N} - \bar{x} \leq - s(N-1)^{-1/2} \tag{3.60}$$

and

$$x_{N:N} - \bar{x} \geq s(N-1)^{-1/2}. \tag{3.61}$$

Proof: (Boyd (1971)). Without loss of generality $\bar{x} = 0$ and $s^2 = 1$. Some of the x_i's are negative say ℓ of them, the remainder are non-negative. Suppose that $x_{1:N} > - 1/\sqrt{N - 1}$ then

$$- \ell/\sqrt{N - 1} < x_{1:N} + \ldots + x_{\ell:N}$$
$$= - (x_{\ell+1:N} + \ldots + x_{N:N})$$

since $\bar{x} = 0$. Thus

$$\sum_{i=1}^{N} x_{i:N}^2 \leq \sum_{i=1}^{\ell} x_{i:N}^2 + 2 \sum_{\ell<i<j\leq N} x_{i:N}x_{j:N} + \sum_{i=\ell+1}^{N} x_{i:N}^2$$
$$= \sum_{i=1}^{\ell} x_{i:N}^2 + \left[\sum_{i=\ell+1}^{N} x_{i:N}\right]^2$$
$$< \ell/(N-1) + \ell^2/(N-1) = \frac{\ell(\ell+1)}{N(N-1)} \leq 1$$

which contradicts $s^2 = 1$. Thus we know $x_{1:N} \leq - 1/\sqrt{N - 1}$. This bound is obtained for the population $x_1 = \ldots = x_{N-1} = - (N-1)^{-1/2}$ and $x_N = (N-1)^{1/2}$. This verifies (3.60). Equation (3.61) is then obtainable by considering $y_i = - x_i$, $i = 1,2,\ldots,N$. An alternative proof is discussed in Exercise 15.

Many other interpretations of (3.57) are possible. Suppose we take a sample of size k from the population of N units $x_1,\ldots,x_N$. We wish to estimate $\bar{x}$ and s based on the sample. Typical estimates are of the form

$$\hat{\bar{x}} = \sum_{i=1}^{k} \lambda_i x_{i:k}, \quad \left[\lambda_i \geq 0, \ \sum_{i=1}^{k} \lambda_i = 1\right] \tag{3.62}$$

and

$$\hat{s} = \left|\sum_{i=1}^{k} \delta_i x_{i:k}\right|, \quad \left[\sum_{i=1}^{k} \delta_i = 0\right] \tag{3.63}$$

where $x_{1:k},\ldots,x_{k:k}$ are the ordered elements in the sample. As a consequence of (3.57) we have

$$\left|\hat{\bar{x}} - \bar{x}\right| \leq s\left[N \sum_{i=1}^{k} \lambda_i^2 - 1\right]^{1/2} \tag{3.64}$$

and

$$\hat{s}/s \leq \left[N \sum_{i=1}^{k} \delta_i^2\right]^{1/2}. \tag{3.65}$$

Differences between order statistics were discussed by Fahmy and Proschan (1981).

Theorem 3.10: For $1 \leq i < j \leq N$,

$$x_{j:N} - x_{i:N} \leq s\sqrt{\frac{(N(N-j+1+i)}{i(N-j+1)}}\,. \tag{3.66}$$

These inequalities are tight. Equality holds for example if $x_1 = \ldots = x_i = 0$, $x_{i+1} = \ldots = x_{j-1} = (N-j+1)/(N-j+1+i)$ and $x_j = \ldots = x_N = 1$.

Proof: Obviously

$$\begin{aligned} x_{j:N} - x_{i:N} &= (x_{j:N}-\bar{x}) - (x_{i:N}-\bar{x}) \\ &\leq \frac{1}{N-j+1} \sum_{k=j}^{N} (x_{k:N}-\bar{x}) - \frac{1}{i} \sum_{k=1}^{i} (x_{k:N}-\bar{x}). \end{aligned}$$

If we apply (3.57) with $\lambda_1 = \ldots = \lambda_i = -\frac{1}{i}$, $\lambda_{i+1} = \ldots = \lambda_{j-1} = 0$ and $\lambda_j = \lambda_{j+1} = \ldots = \lambda_N = \frac{1}{N-j+1}$, (3.66) follows. Fahmy and Proschan (1981) provide an alternative proof based on arguments involving the nature of the extremal population. David, Hartley and Pearson (1954) derived (3.66) in the special case $i = 1$, $j = N - 1$ (their s^2 is $N/(N-1)$ times the s^2 used in the present discussion).

Special cases of Theorem 3.10 yield bounds on the range

$$x_{N:N} - x_{1:N} \leq s\sqrt{2N} \tag{3.67}$$

(already noted following equation (3.55) and originally due to Nair (1948) and Thomson (1955)), on quasi-ranges,

$$x_{N-k+1:N} - x_{k:N} \leq s\sqrt{2N/k} \tag{3.68}$$

and on spacings

$$x_{k+1:N} - x_{k:N} \leq sN/\sqrt{k(N-k)}. \tag{3.69}$$

As Fahmy and Proschan observed, it is not possible to give non-trivial lower bounds for $x_{j:N} - x_{i:N}$, except in the case where $j = N$ and $i = 1$ (i.e. the case of the range). In that case one has the bound

$$\begin{aligned} x_{N:N} - x_{1:N} &\geq 2s \quad (N \text{ even}) \\ &\geq 2s(1-N^{-2})^{-1/2} \quad (N \text{ odd}) \end{aligned} \tag{3.70}$$

a result previously noted by Thomson (1955). See Exercises 11 and 16.

Theorem 3.8 and 3.9 can be interpreted as giving bounds on $x_{j:N}$ subject to the constraints that $\sum_{i=1}^{N} x_i = 0$ and $\sum_{i=1}^{N} x_i^2 = 1$. Beesack (1973) determined analogous results replacing the constraint $\sum_{i=1}^{N} x_i^2 = 1$ by a more general one.

Theorem 3.11: (Beesack (1973)). Let $x_1,\dots,x_N$ satisfy $\sum_{i=1}^{N} x_i = 0$ and $\sum_{i=1}^{N} f(|x_i|) = 1$ where f is non-negative, strictly increasing and convex on $[0,\infty)$ with $f(0) = 0$ and $f(x) > 1$ for some $x > 1$. It follows that

$$-\alpha_1 \le x_{1:N} - \beta_1 \tag{3.71}$$

$$-\alpha_j \le x_{j:N} \le \alpha_{N-j+1}, \quad j = 2,\dots,N-1 \tag{3.72}$$

and

$$\alpha_N \le x_{N:N} \le \beta_N \tag{3.73}$$

where α_j $(j = 1,2,\dots,N-1)$ is the unique positive solution of the equation

$$jf(x) + (N-j)f\left[\frac{jx}{N-j}\right] = 1,$$

$\beta_N = \alpha_1$, $\beta_1 = \alpha_N$ and α_N is the unique positive solution of the equation

$$(N-1)f(x) + f((N-1)x) = 1.$$

The bounds (3.71) – (3.73) are best possible.

In particular if we take $f(x) = x^p$ where $p \ge 1$, the bounds take the form

$$-\left[\frac{(N-1)^{p-1}}{1 + (N-1)^{p-1}}\right]^{1/p} \le x_{1:N} \le -\left[\frac{1}{(N-1)^p + N - 1}\right]^{1/p}, \tag{3.74}$$

$$-\left[\frac{(N-j)^{p-1}}{j^p + j(N-j)^{p-1}}\right]^{1/p} \le x_{j:N} \le \left[\frac{(j-1)^{p-1}}{(N-j+1)^p + (N-j+1)(j-1)^{p-1}}\right]^{1/p}, \tag{3.75}$$

$$\left[\frac{1}{(N-1)^p + (N-1)}\right]^{1/p} \le x_{N:N} \le \left[\frac{(N-1)^{p-1}}{1 + (N-1)^{p-1}}\right]^{1/p}. \tag{3.76}$$

The case $p = 2$ corresponds to the results in Theorems 3.8 and 3.9. Beesack also obtained bounds for the case $f(x) = x^p$ where $p < 1$ (see Exercise 14). The case $p = 1$ corresponds to bounds in mean deviation units. Thus

Corollary 3.12: If we define $s' = \frac{1}{N}\sum_{i=1}^{N} |x_i - \bar{x}|$ then

$$-\frac{N}{2}s' \le x_{1:N} - \bar{x} \le -\frac{N}{2(N-1)}s', \tag{3.77}$$

$$-\frac{N}{2j}s' \le x_{j:N} - \bar{x} \le \frac{N}{2(N-j+1)}s', \quad j = 2,\dots,N-1 \tag{3.78}$$

and

$$\frac{N}{2(N-1)}s' \le x_{N:N} - \bar{x} \le \frac{N}{2}s'. \tag{3.79}$$

These bounds are best possible.

Arnold and Groeneveld (1981) give mean deviation bounds in a finite population sampling context. If we denote the mean of a sample of size n by $\bar{x}^{(n)}$ and s' as in Corollary 3.12, their result may be written as

Theorem 3.13:

$$|\bar{x}^{(n)} - \bar{x}| \le s'N/(2n). \tag{3.80}$$

Proof: Without loss of generality $\bar{x} = 0$ and $Ns' = 2 \sum_{x_i > 0} x_i$. Multiply (3.80) by n and we have $|n\bar{x}^{(n)}| \le \sum_{x_i > 0} x_i$ which is clearly true.

Bounds in range and mean units are also provided in Arnold and Groeneveld (1981).

Theorem 3.14:

$$|\bar{x}^{(n)} - \bar{x}| \le [1 - (n/N)][x_{N:N} - x_{1:N}]. \tag{3.81}$$

Theorem 3.15: If the x_i's are positive then

$$|\bar{x}^{(n)} - \bar{x}| \le \bar{x} \max\{1, (N/n) - 1\}. \tag{3.82}$$

The last result includes a bound derived by Koop (1972). Proofs of these theorems together with those of analogous results for symmetric populations are assigned to Exercises 19 and 20.

Analogs of Corollary 3.12 using range and mean units were provided by Groeneveld (1982).

Theorem 3.16:

$$x_{1:N} - \bar{x} \le -[x_{N:N} - x_{1:N}]/N, \tag{3.83}$$

$$-[x_{N:N} - x_{1:N}](N-k)/N \le x_{k:N} - \bar{x} \le [x_{N:N} - x_{1:N}](k-1)/N, \quad k = 2,3,\ldots,N-1 \tag{3.84}$$

and

$$x_{N:N} - \bar{x} \ge [x_{N:N} - x_{1:N}]/N. \tag{3.85}$$

Proof: Without loss of generality $x_{N:N} = 1$ and $x_{1:N} = 0$. The extremal populations are readily identified. In all cases they consist of N values some of which are 0 and the rest of which are 1.

Theorem 3.17: If the x_i's are non–negative then for $k = 1,2,\ldots,N$

$$0 \le x_{k:N} \le N\bar{x}/(N-k+1). \tag{3.86}$$

Proof: The extremal populations can be as described in the proof of Theorem 3.16.

Throughout this section we have considered N real numbers $x_1,x_2,\ldots,x_N$ and the bounds obtained have been sure bounds. If we consider N random variables $X_1,X_2,\ldots,X_N$ then almost sure bounds are obtainable for their realized values. Thus, for example,

$$X_{n:n} - \bar{X} \le S\sqrt{N-1} \quad \text{a.s.} \tag{3.87}$$

where $S^2 = \frac{1}{N}\sum_{i=1}^{N}(X_i - \bar{X})^2$, using Theorem 3.3 applied to all realized values of $X_1,\ldots,X_n$. Note that the X_i's in (3.87) do not have to be independent, nor identically distributed, merely all defined on the same space. We will use the following notation for expectations:

$$\mu_i = E(X_i), \quad i = 1,2,\ldots,N$$

$$\mu_{i:N} = E(X_{i:N}), \quad i = 1,2,\ldots,N$$

$$\sigma_i^2 = \text{var}(X_i), \quad i = 1,2,\ldots,N \tag{3.88}$$

and

$$\bar{\mu} = \frac{1}{N}\sum_{i=1}^{N} \mu_i = E(\bar{X}).$$

From (3.87) we have

$$\mu_{n:n} - \bar{\mu} \le \sqrt{N-1}\, E(S). \tag{3.89}$$

The bound in (3.89) is inconvenient since we cannot express E(S) in terms of means and variances. Note that

$$E(S) \le \sqrt{E(S^2)} \tag{3.90}$$

and

$$\begin{aligned} NE(S^2) &= E\left[\sum_{i=1}^{N} (X_i - \overline{X})^2\right] \\ &= E\left[\sum_{i=1}^{N} X_i^2\right] - NE(\overline{X}^2) \\ &\le \sum_{i=1}^{N} E(X_i^2) - N[E(\overline{X})]^2 \qquad (3.91) \\ &= \sum_{i=1}^{N} (\sigma_i^2 + \mu_i^2) - N\overline{\mu}^2 \\ &= \sum_{i=1}^{N} [\sigma_i^2 + (\mu_i - \overline{\mu})^2]. \end{aligned}$$

We thus have

$$\mu_{n:n} - \overline{\mu} \le [(N-1)/N]^{1/2} \sqrt{\sum_{i=1}^{N} \left[\sigma_i^2 + (\mu_i - \overline{\mu})^2\right]}. \tag{3.92}$$

The bound is simplified under homogeneity assumptions $\mu_i = \mu \; \forall \; i$ and $\sigma_i^2 = \sigma^2 \; \forall \; i$. In that case

$$\mu_{n:n} \le \mu + \sigma\sqrt{N-1}. \tag{3.93}$$

Note that we did not assume independence nor identical distributions in the derivation of (3.93).

Many of the earlier results easily yield bounds on expectations of linear functions of order statistics. We have

Theorem 3.18: (Arnold and Groeneveld (1979), Nagaraja (1981)). Let $X_1, X_2, \ldots, X_N$ be jointly distributed random variables with means and variances given by (3.88), then for any $\underline{\lambda} \in \mathbb{R}^N$

$$\begin{aligned} \left|\sum_{i=1}^{N} \lambda_i(\mu_{i:N} - \overline{\mu})\right| &\le E(S)\sqrt{N\left[\sum_{i=1}^{N} (\lambda_i - \overline{\lambda})^2\right]} \qquad (3.94) \\ &\le \sqrt{NE(S^2)\sum_{i=1}^{N} (\lambda_i - \overline{\lambda})^2} \le \left[\sum_{i=1}^{N} \left[\sigma_i^2 + (\mu_i - \overline{\mu})^2\right] \sum_{i=1}^{N} (\lambda_i - \overline{\lambda})^2\right]^{1/2}. \end{aligned}$$

In particular, if $\mu_i - \mu \; \forall \; i$ and $\sigma_i^2 = \sigma^2 \; \forall \; i$,

$$\left|\sum_{i=1}^{N} \lambda_i(\mu_{i:N} - \mu)\right| \le \sigma\sqrt{N\sum_{i=1}^{N} (\lambda_i - \overline{\lambda})^2}. \tag{3.95}$$

Proof: For each realization $x_1, x_2, \ldots, x_N$ we have (3.57). Taking expectations we obtain the first inequality in (3.94). The other inequalities follow from (3.90) and (3.91). Equation (3.95) is obtained by substitution in (3.94).

The analog of Theorem 3.8 holds

Theorem 3.19: Let $X_1,X_2,\ldots,X_N$ be jointly distributed random variables with means and variances given by (3.88). For i = 1,2,...,N

$$
\begin{aligned}
-\sqrt{E(S^2)}\sqrt{\frac{N-i}{N}} &\le E(S)\sqrt{\frac{N-i}{N}} \\
&\le \mu_{i:N} - \bar{\mu} \qquad (3.96)\\
&\le E(S)\sqrt{\frac{i-1}{N-i+1}} \le \sqrt{E(S^2)}\sqrt{\frac{i-1}{N-i+1}}
\end{aligned}
$$

where $E(S^2)$ may be replaced by the bound given in (3.91). In particular if $\mu_i = \mu\ \forall\ i$ and $\sigma_i^2 = \sigma^2\ \forall\ i$ we have

$$-\sigma\sqrt{\frac{N-i}{i}} \le \mu_{i:N} - \mu \le \sigma\sqrt{\frac{i-1}{N-i+1}}. \qquad (3.97)$$

Proof: Use (3.58).

Note that the bounds (3.96) and (3.97) are tight. They are achieved for example by a random vector $(X_1,\ldots,X_N)$ which denotes an exhaustive sample drawn without replacement from a population of the type discussed in Exercise 9. Analogs of Theorems 3.10, 3.13, 3.14, 3.15, 3.17 may be obtained by considering random X_i's and taking expectations. Similar comments apply to many of the Exercises at the end of the chapter. A caveat is in order however. Replacement of E(S) by $\sqrt{E(S^2)}$ is not always appropriate. Nagaraja (1979) points out that care must be taken for example with extension of Theorem 3.9. Theorem 3.9 implies that

$$X_{1:n} - \bar{X} \le -S(N-1)^{-1/2} \text{ a.s.} \qquad (3.98)$$

We may legitimately take expectations here obtaining

$$\mu_{1:n} - \bar{\mu} \le -E(S)(N-1)^{-1/2}. \qquad (3.99)$$

Since there is a minus sign on the upper bound in (3.99) it is not legitimate to replace E(S) by $\sqrt{E(S^2)}$ (see Exercise 31).

Many of the bounds in this section can be improved if knowledge of the covariances between the X_i's is available. See for example, Exercises 59 and 60.

3.3. Bounds via maximal dependence

In Section 2 several bounds were obtained for expectations of functions of order statistics from possibly dependent samples. If we focus on extreme order statistics we can derive alternative proofs and some new results by using the concept of maximal dependence introduced by Lai and Robbins (1976).

Let $X_1,X_2,\ldots,X_n$ be jointly distributed possibly dependent random variables with corresponding marginal distribution functions $F_1,F_2,\ldots,F_n$. As usual denote the largest of the X_i's by $X_{n:n}$. The Bonferroni inequalities immediately yield bounds on the distribution function of $X_{n:n}$. Thus for every real x

$$\max\left\{0,1-\sum_{i=1}^{n} F_i(x)\right\} \le F_{X_{n:n}}(x) \le \min_i F_i(x). \tag{3.100}$$

Both of these bounds are achievable. A set of random variables $X_1,\ldots,X_n$ for which the right hand inequality is achieved may be constructed as follows (F_i^{-1} is as defined in equation (1.3)). Let U be a single uniform (0,1) random variable and, for i = 1,2,...,n, define

$$X_i = F_i^{-1}(U). \tag{3.101}$$

Evidently with this definition

$$\begin{aligned} P(X_{n:n} \le x) &= P(X_i \le x, \forall\, i) \\ &= P(U \le F_i(x), \forall\, i) \\ &= \min_i F_i(x). \end{aligned}$$

The achievability of the left hand inequality is less self-evident. Lai and Robbins (1975) first verified this possibility and called a random vector $X_1,X_2,\ldots,X_n$ maximally dependent if it achieved the bound. Their work builds on an example with uniform marginals presented by Mallows. Lai and Robbins (1978) discuss construction of maximally dependent sequences. A direct construction of a random vector achieving the left hand bound in (3.100) is possible as follows.

Let $x_0 = \inf\left\{x: \sum_{i=1}^{n} F_i(x) \le 1\right\}$, such an x_0 always exists (finite) since $F_i(x) \uparrow 1$ as $x \to \infty$ for every i. Consider the unit interval $\Omega = \{\omega:\ 0 \le \omega \le 1\}$ with Lebesgue measure as a probability space and define random variables $Z_1,Z_2,\ldots,Z_n$ on the space as follows. For i = 1,2,...,n

$$\begin{aligned} z_i(\omega) &= \omega - c_i + 1, \quad 0 \le \omega < c_i \\ &= \omega - c_i, \quad c_i \le \omega \le 1 \end{aligned} \tag{3.102}$$

where

$$c_i = \sum_{j<i} F_j(x_0), \quad i = 1,2,\ldots,n+1.$$

If one draws the graphs of the z_i's it becomes evident that they are uniformly distributed (remember that c_{n+1} may be strictly less than 1). It is convenient to divide the sample space $\Omega = [0,1]$ into n+1 disjoint subsets, namely the intervals $[c_1,c_2),[c_2,c_3),\ldots,[c_{n+1},1]$ which we denote by $\Omega_1,\Omega_2,\ldots,\Omega_{n+1}$. Now define random variables $X_1,X_2,\ldots,X_n$ on Ω by

$$X_i = F_i^{-1}(Z_i), \quad i = 1,2,\ldots,n. \tag{3.103}$$

We claim that $X_{n:n} = \max_i X_i$ has as its distribution function, the left hand bound in (3.100), i.e. $\underline{X}$ is maximally dependent. This follows since (here it may help to refer to a diagram of the case n = 3), for x $\ge x_0$

$$\begin{aligned} P(X_{n:n} > x) &= P\left[\bigcup_{i=1}^{n} \{X_i > x\}\right] \\ &= P\left[\bigcup_{i=1}^{n} \{Z_i > F_i(x)\}\right]. \end{aligned}$$

By construction the events $\{Z_i > F_i(x)\}$ are disjoint ($\{Z_1 > F_1(x)\} \subset \Omega_n \cup \Omega_{n+1}$ while $\{Z_i > F_i(x)\}$ $\subset \Omega_{i-1}$, i = 2,3,...,n). Consequently for $x \ge x_0$,

$$P(X_{n:n} > x) = \sum_{i=1}^{n} P\left[Z_i > F_i(x)\right]$$

$$= \sum_{i=1}^{n} \bar{F}_i(x).$$

Thus $\underline{X}$ is maximally dependent.

Having a bound on the distribution of $X_{n:n}$ we may immediately obtain a bound on its expectation provided $E(X_i^+)$ exists for every i.

Theorem 3.20: (Lai–Robbins (1976)). If $E(X_i^+) < \infty$, $i = 1,2,\ldots,n$ then

$$E(X_{n:n}) \le x_0 + \sum_{i=1}^{n} \int_{x_0}^{\infty} \bar{F}_i(y)dy, \tag{3.104}$$

where $x_0 = \inf\left\{x: \sum_{i=1}^{n} \bar{F}_i(x) \le 1\right\}$.

Proof: Since $(X_{n:n} - x_0)^+$ is a non–negative random variable whose expectation can be obtained by integrating its survival function, the result follows from the upper bound for $\bar{F}_{X_{n:n}}(x)$ provided by the left hand inequality in (3.100).

Corollary 3.2.1: (Arnold (1980)). If $F_i \equiv F$, $i = 1,2,\ldots,n$ and $E(X_1^+) < \infty$ then

$$E(X_{n:n}) \le n \int_{(1-n^{-1})}^{1} F^{-1}(u)du. \tag{3.105}$$

Proof: Clearly $x_0 = F^{-1}(1 - n^{-1})$. The result then follows from (3.104) by an application of Fubini's theorem and a convenient change of variables.

Bound for $E(X_{j:n})$ ($j \ne 1$ or n) analogous to (3.105) have been sought. Gravey (1985) gives some results assuming exchangeability of the X_i's.

Corollary 3.2.1 can be used to derive universal bounds on $E(X_{n:n})$ in a manner analogous to that used by Gumbel (1954) and Hartley and David (1954). The scenario involves n random variables $X_1,X_2,\ldots,X_n$ possibly dependent but with common marginal distribution F.

Theorem 3.2.2: Suppose $X_1,X_2,\ldots,X_n$ are identically distributed, possibly dependent with $E(X_1) = \mu$ and $\mathrm{var}(X_i) = \sigma^2$, assumed finite. It follows that

$$E(X_{n:n}) \le \mu + \sigma\sqrt{n-1}. \tag{3.106}$$

The inequality is tight and is achieved by $X_1,\ldots,X_n$ maximally dependent in the sense of Lai and Robbins (1976) with common distribution of the form

$$P\left[X_i = \mu - \sigma(n-1)^{-1/2}\right] = 1 - n^{-1}$$

and

$$P\left[X_i = \mu + \sigma(n-1)^{1/2}\right] = n^{-1}. \tag{3.107}$$

Proof: Without loss of generality $\mu = 0$, $\sigma^2 = 1$, i.e. $\int_0^1 F^{-1}(u)du = 0$, $\int_0^1 [F^{-1}(u)]^2du = 1$. Using (3.105), the Schwarz inequality and a function $g(u) = nI(u \ge 1 - n^{-1})$ we have

$$E(X_{n:n}) \le n \int_{(1-n^{-1})}^{1} F^{-1}(u)du$$
$$= \int_0^1 (g-1)F^{-1}du \quad \left[\text{since} \int_0^1 F^{-1}du = 0\right]$$
$$\le \left\{\int_0^1 (g-1)^2du \int_0^1 (F^{-1})^2du\right\}^{1/2}$$
$$= \sqrt{n-1} \quad \left[\text{since} \int_0^1 (F^{-1})^2\, du = 1\right].$$

Equality obtains if $F^{-1} \propto g-1$. It is readily verified that (with $\mu = 0$, $\sigma = 1$) the distribution determined by (3.107) has its inverse proportional to $g-1$.

Although the proof of Theorem 3.22 is attractive, it should be remarked that the result in question is a special case of equation (3.93) earlier derived (without assuming identical distributions for the X_i's, just common mean μ and variance σ^2). Corollary 3.21 does give us some results not easily derived using the techniques of Section 2. A bound for $E(X_{n:n})$ assuming a common symmetric distribution for the X's is obtained as follows.

Theorem 3.23: Suppose $X_1, X_2, \ldots, X_n$ are identically distributed with a common symmetric distribution with mean μ and variance σ^2 (assumed finite). It follows that

$$E(X_{n:n}) \le \mu + \sigma\sqrt{(n/2)}. \tag{3.108}$$

The inequality is tight. It is achieved by $X_1, X_2, \ldots, X_n$ maximally dependent in the sense of Lai and Robbins (1976) with common distribution of the form

$$P(X_i = \mu + \sigma\sqrt{(n/2)}\,) = P(X_i = \mu - \sigma\sqrt{(n/2)}\,) = n^{-1} \tag{3.109}$$

and

$$P(X_i = \mu) = 1 - 2n^{-1}.$$

Proof: Assume $\mu = 0$, $\sigma^2 = 1$. Using (3.105) we seek to maximize $n \int_{(1-n^{-1})}^{1} F^{-1}(u)du$ subject to $\int_{1/2}^{1} [F^{-1}(u)]^2du = \frac{1}{2}$ [by symmetry for $u < \frac{1}{2}$ we have $F^{-1}(u) = -F^{-1}(1-u)$]. Again let $g(u) = nI(u \ge 1 - n^{-1})$ and obtain using the Schwarz inequality

$$E(X_{n:n}) \le \int_{1/2}^{1} gF^{-1}(u)du$$
$$\le \left\{\int_{1/2}^{1} g^2du \int_{1/2}^{1} (F^{-1})^2du\right\}^{1/2}$$
$$= \sqrt{(n/2)}.$$

For tightness (with $\mu = 0$, $\sigma^2 = 1$) it is readily verified that the inverse distribution function corresponding to (3.109) is proportional to g over the interval $(\frac{1}{2}, 1)$.

The result of this theorem is not surprising in the light of Exercise 13 (set n = 1, in that Exercise). One cannot just take expectations in Exercise 13. Although the X_i's are symmetric r.v.'s the realized vector $X_1, X_2, \ldots, X_n$ may not (indeed probably will not) be symmetric about $\bar{X}$.

Arnold (1980) provided tables comparing the four bounds obtained for $E(X_{n:n})$, namely equations (3.4) (independent X_i's), (3.11) (independent symmetric X_i's), (3.106) (possibly dependent X_i's) and (3.108) (possibly dependent symmetric X_i's). For large n there is little difference between the effect of assuming symmetry alone and the effect of assuming independence alone (since $\sqrt{(n/2)} \sim (n-1)(2n-1)^{-1/2}$).

If we wish to consider the expected range of $X_1,X_2,\ldots,X_n$ possibly dependent with common marginal distribution F, we may apply (3.105) to $X_1,\ldots,X_n$ and $-X_1,\ldots,-X_n$ obtaining

$$E(X_{n:n} - X_{1:n}) \le 2\left[\int_{(1-n^{-1})}^{1} F^{-1}(u)du - \int_{0}^{n^{-1}} F^{-1}(u)du\right]. \tag{3.110}$$

For example if the X_i's are uniform (0,1) we conclude that the expected range is never larger than $(n-1)/n$ (an achievable bound (see Exercise 40)). Using (3.110) and the Schwarz inequality we readily deduce

Theorem 3.24: If $X_1,X_2,\ldots,X_n$ are identically distributed with common distribution having mean μ and variance σ^2 then

$$E(X_{n:n} - X_{1:n}) \le \sigma\sqrt{2n}\,. \tag{3.111}$$

This bound is tight.

We omit the proof of (3.111). The more general result assuming possibly different distributions for the X_i's with common mean μ and variance σ^2 can be obtained by taking expectations in the Nair–Thomson bound for the range (3.56) using (3.90) and (3.91) (with $\mu_i = \mu$, $\sigma_i^2 = \sigma^2$).

Gilstein (1981) was the first to remark on the possibility of using the Holder inequality instead of the Schwarz inequality in deriving bounds on $E(X_{n:n})$. The arguments are essentially unchanged. A survey of such results is provided by Arnold (1985) (see Exercises 41–47).

3.4. Restricted families of parent distributions

Naturally we can expect tighter bounds on moments of order statistics if we impose restrictions on the class of possible parent distributions. The effect of symmetry, for example, has been discussed in earlier sections.

Blom (1958) observes that if we write

$$E(X_{i:n}) = \int_0^1 F^{-1}(u)\, g_{i:n}(u)du \tag{3.112}$$

where $g_{i:n}$ is the density of the i'th order statistic from a uniform (0,1) random sample, and if F^{-1} is convex then Jensens inequality may be applied $\left[\text{recall } E(U_{i:n}) = \frac{i}{n+1}\right]$. Thus

Theorem 3.25: (Blom). If $X_1,X_2,\ldots,X_n$ are i.i.d. with common distribution having a convex inverse then

$$E(X_{i:n}) \le F^{-1}(i/(n+1)). \tag{3.113}$$

If F^{-1} is concave then

$$E(X_{i:n}) \ge F^{-1}(i/(n+1)). \tag{3.114}$$

A direct extension of Blom's result is possible. Consider two distribution functions F and G such that $F^{-1}G$ is convex on the support of G (Blom considered the case where $G(x) = x$). If $Y \sim G$ then $F^{-1}G(Y)$

~ F. Directly from Jensen's inequality we get

Theorem 3.26: If $X_1,X_2,\ldots,X_n$ are i.i.d. F and $Y_1,Y_2,\ldots,Y_n$ are i.i.d. G where $F^{-1}G$ is convex on the support of G then

$$G(E(Y_{i:n})) \leq F(E(X_{i:n})). \tag{3.115}$$

Proof:

$$E(X_{i:n}) = E[F^{-1}G(Y_{i:n})]$$
$$\geq F^{-1}G\, E(Y_{i:n}).$$

If F has support $[0,\infty)$ we can apply Jensens inequality twice. The following theorem is implicit in Barlow and Proschan (1966) (see also Nagaraja (1981)).

Theorem 3.27: Suppose $X_i \geq 0$, $i = 1,2,\ldots,n$ are i.i.d. F and Y_i, $i = 1,2,\ldots,n$ are i.i.d. G where $F^{-1}G$ is convex on the support of G. Then for any $\underline{\lambda} = (\lambda_1,\ldots,\lambda_n)$ with $\lambda_i \geq 0$, $\sum_{i=1}^{n} \lambda_i = 1$ we have

$$G\left[\sum_{i=1}^{n} \lambda_i E(Y_{i:n})\right] \leq F\left[\sum_{i=1}^{n} \lambda_i E(X_{i:n})\right]. \tag{3.116}$$

Proof:

$$\sum_{i=1}^{n} \lambda_i E(X_{i:n}) = \sum_{i=1}^{n} \lambda_i E[F^{-1}G(Y_{i:n})]$$
$$\geq \sum_{i=1}^{n} \lambda_i F^{-1}G[E(Y_{i:n})] \quad \text{(Jensen and non-negativity)}$$
$$\geq F^{-1}G\left[\sum_{i=1}^{n} \lambda_i E(Y_{i:n})\right] \quad \text{(Jensen).}$$

A distribution function G with support $[0,\infty)$ has increasing failure rate (IFR) if and only if $F^{*-1}G$ is convex where $F^*(x) = 1 - e^{-x}$ (the standard exponential distribution). The expected values of the corresponding exponential order statistics are given by

$$\mu^*_{i:n} = \sum_{j=1}^{i} \frac{1}{n-j+1}. \tag{3.117}$$

We may then enunciate

Corollary 3.27: Suppose $Y_i \geq 0$, $i = 1,2,\ldots,n$ are i.i.d. random variables whose common distribution G has increasing failure rate. For any $\underline{\lambda} = (\lambda_1,\ldots,\lambda_n)$ with $\lambda_i \geq 0$, $\sum_{i=1}^{n} \lambda_i = 1$ we have

$$G\left[\sum_{i=1}^{n} \lambda_i E(Y_{i:n})\right] \leq 1 - \exp\left[-\sum_{i=1}^{n} \lambda_i \sum_{j=1}^{i} \frac{1}{n-j+1}\right]. \tag{3.118}$$

In particular

$$G(E(Y_{i:n})) \leq 1 - \exp\left[-\sum_{j=1}^{i} \frac{1}{n-j+1}\right]. \tag{3.119}$$

Note that if G is DFR the inequalities in (3.118) and (3.119) are reversed.

Ali and Chan (1965) considered unimodal and U–shaped distributions, extending Blom's result in another direction. We say that F is unimodal if there exists c such that F is convex on $(-\infty,c)$ and concave on (c,∞). Ali and Chan focus on symmetric unimodal distributions. If we assume without loss of generality that they are to be symmetric about zero then they are dealing with symmetric distributions whose inverse F^{-1} is convex on $(\frac{1}{2}, 1)$ and they are comparing the expectations of order statistics of F to those corresponding to a uniform distribution. Thus we are really dealing with s–comparisons in the sense of Van Zwet (1964). The more general result (essentially provided by Nagaraja (1981)) is as follows.

Theorem 3.28: Let X_i, $i = 1,2,\ldots,n$ be i.i.d. F and Y_i, $i = 1,2,\ldots,n$ be i.i.d. G where F and G are symmetric about 0 and $F^{-1}G$ is convex on the positive part of the support of G. It follows that for any $\underline{\lambda} = (\lambda_1,\ldots,\lambda_n)$ with $\lambda_i = 0$, $i < (n+1)/2$ and $\lambda_i \geq 0$, $i \geq (n+1)/2$ such that $\sum_{i=1}^{n} \lambda_i = 1$ we have

$$G\left[\sum_{i=1}^{n} \lambda_i E(Y_{i:n})\right] \leq F\left[\sum_{i=1}^{n} \lambda_i E(X_{i:n})\right] \tag{3.120}$$

Proof: As in Theorem 3.27 using the fact that for $i \geq (n+1)/2$ we have $E(Y_{i:n}) \geq 0$.

Corollary 3.29: (Ali and Chan (1965)). If $X_1,X_2,\ldots,X_n$ are i.i.d. F where F is symmetric about 0 and unimodal, then for $i \geq (n+1)/2$ we have

$$E(X_{i:n}) \geq F^{-1}(i/(n+1)). \tag{3.121}$$

Proof: Apply Theorem 3.28 with $G(x) = x$ and $\underline{\lambda} = (0,\ldots,1,\ldots,0)$ with a 1 in the i'th coordinate.

A distribution function F that is symmetric about 0 is said to be U–shaped (following Ali and Chan (1965)) if F^{-1} is concave on $(\frac{1}{2}, 1)$. It follows that for G corresponding to a uniform (0,1) distribution we have $G^{-1}F$ convex on the positive support of F. Theorem 3.28 applies (with the roles of F and G reversed) and yields

Corollary 3.30: (Ali and Chan (1965)). If $X_1,X_2,\ldots,X_n$ are i.i.d. F where F is symmetric about 0 and U–shaped, then for $i \geq (n+1)/2$ we have

$$E(X_{i:n}) \leq F^{-1}(i/(n+1)). \tag{3.122}$$

Barlow and Proschan (1966) describe certain inequalities for expectations of order statistics based on star ordering. Attention is restricted to distributions with support $[0,\infty)$. A function ϕ: $[0,\infty) \to (-\infty,\infty)$ is said to be star–shaped if $\phi(\alpha x) \leq \alpha\phi(x)$, $\forall\, x \geq 0$, $\forall\, \alpha \in [0,1]$. Convex functions are examples of star–shaped functions but non–convex examples exist (Exercise 49). We saw earlier that it was possible to relate expectations of order statistics from F and G when $F^{-1}G$ was convex. Not surprisingly the condition that $F^{-1}G$ be star–shaped also yields certain inequalities. A useful result in this context is the following.

Lemma 3.31: (Barlow and Proschan (1966)). Let $h_{i:n}$ denote the density of the i'th order statistic from a sample of size n from a uniform (0,1) distribution. Suppose that g: $[0,1] \to R$ changes sign from + to – exactly once in the interval [0,1]. It follows that (provided the relevant integrals converge)

$$a_{i:n} = \int_0^1 g(u)h_{i:n}(u)du$$

changes sign at most once (from + to –) as i increases from 1 to n for fixed n. Similarly for rixed i, $a_{i:n}$ changes sign at most once (from – to +) as n increases from i to ∞.

Proof: Omitted. The result is certainly plausible since for large n, the density $h_{i:n}$ is highly concentrated in a neighborhood of $u \approx i/n$. A careful proof involves variation diminishing properties of totally positive functions (cf. Karlin (1968)).

Lemma 3.31 provides us immediately with

Theorem 3.32: (Barlow and Proschan (1966)). Let $X_i \geq 0$, $i = 1,2,...$ be i.i.d. F and $Y_i \geq 0$, $i = 1,2,...$ be i.i.d. G where $F^{-1}G$ is star-shaped on the support of G. It follows that $E(X_{i:n})/E(Y_{i:n})$ is

(i) increasing in i

and (ii) decreasing in n.

Proof: If ϕ is star-shaped then for any c, $u - c\phi(u)$ changes sign at most once (from + to −). Since $F^{-1}G$ is star-shaped it follows that $G^{-1}(u) - cF^{-1}(u)$ changes sign (+,−) at most once. However

$$E(Y_{i:n}) - cE(X_{i:n}) = \int_0^1 [G^{-1}(u) - cF^{-1}(u)]h_{i:n}(u)du$$

and so by Lemma 3.31 $[E(Y_{i:n})/E(X_{i:n})] - c$ changes sign (+,−) at most once as i increases. Since this is true for every c, the ratio $E(Y_{i:n})/E(X_{i:n})$ decreases as i increases. Result (i) follows. Result (ii) is similarly verified.

Special cases of Theorem 3.32 of practical interest include: (a) The case $F(x) = x$, in which case G^{-1} should be star-shaped. The conclusion of the theorem becomes: $(n+1)E(Y_{i:n})/i$ is decreasing in i and increasing in n. (b) The case $F(x) = 1 - e^{-x}$, in which case G is an increasing failure rate average (IFRA) distribution. The conclusion is that $E(Y_{i:n})/\left[\sum_{j=1}^{i} (n-j+1)^{-1}\right]$ is decreasing in i and increasing in n.

Exercises

1. Consider what Gumbel calls the expected largest value of the distribution (3.7). It is that value of x, say x_n, for which $F(x_n) = (n-1)/n$. Give a simple asymptotic expression for x_n.

2. Verify that, for large values of n, the bounds (3.4) and (3.11) are well approximated by $\sqrt{(n-1)/2}$ and $\sqrt{n+1/2}\,/2$ respectively.

3. Suppose $X_1, X_2, \ldots, X_n$ are i.i.d. with $E(X_i) = 0$, $\operatorname{var}(X_i) = 1$ and suppose that $X_i \le 1$, $\forall$ i. The upper bound (3.4) is not attainable now. How should it be modified (cf. Hartley and David (1954)).

4. (Beesack (1973)). Let f be continuous and strictly increasing on $[0,\infty)$ with $f(0) = 0$ and $f(x) > A$ for some $x > 0$ where $A > 0$. Suppose that the real numbers $x_1, x_2, \ldots, x_N$ satisfy $\sum_{i=1}^{N} x_i = 0$, $|x_1| < |x_2| < \cdots < |x_N|$ and $\sum_{i=1}^{N} f(|x_i|) = A$. Then $|x_i| \le \alpha_i$ where α_i is the unique positive root of the equation

$$(N-i+1)f(x) = A, \quad 1 \le i \le N.$$

Moreover, if $(N-i+1)$ is even, then the bound α_i is best possible.

5. (Beesack (1973)). In Exercise 4, if in addition f is strictly convex on $[0,\infty)$ and $(N-i+1)$ is odd with $i > 1$, then $|x_i| \le \beta_i$ where β_i is the unique positive root of the equation

$$(i-1)f\left[\frac{x}{i-1}\right] + (N-i+1)f(x) = A, \quad 1 < i \le N.$$

If $i = 1$, N is odd and f is strictly convex then $|x_1| \le \beta_1$ where $\beta_1 = \max\left\{\beta_{ij}\colon \frac{N+1}{2} \le j \le N-1\right\}$ and β_{ij} is the unique positive root of the equation

$$i\,f(x) + (N-i)f\left[\frac{ix}{N-i}\right] = A.$$

These bounds β_i, $i = 1,2,\ldots,N$, are best possible.

6. (Beesack (1973)). Show that the choice $f(x) = x^p$ $(p \ge 1)$ and $A = 1$ in Exercises 4 and 5 yields the bounds

$$|x_i| \le (N-i+1)^{-1/p}, \quad \text{if } N-i+1 \text{ even},$$

$$|x_i| \le [(i-1)^{1-p} + N - i + 1]^{-1/p}, \quad \text{if } N-i+1 \text{ odd and } i \ne 1,$$

and

$$|x_i| \le (N-1)2^{1/p}[(N-1)(N+1)^p + (N+1)(N-1)^p]^{-1/p}, \quad \text{if N is odd.}$$

The choice $p = 2$ leads to Scott's results (Theorem 3.2).

7. Devise a proof of the Samuelson–Scott inequality (3.36) using the arithmetic–geometric mean inequality.

8. In Corollary 3.7, show that if $n_1 + n_2 > N$ then the squared generalized distance between $\bar{x}^{(n_1)}$ and $\bar{x}^{(n_2)}$ cannot exceed $N(2N - n_1 - n_2)/(n_1 n_2)$ (in this case we must have $\ell > 0$). (The one dimensional version of this result is equation (6.2) of Mallows and Richter (1969)).

9. Verify that the bounds exhibited in Theorem 3.8 for $i = 2,\ldots,N-1$ are tight by exhibiting simple populations in which the bounds are achieved.

10. (Mallows and Richter (1969)). Let $x_1,x_2,\ldots,x_N$ be such that $\bar{x} = 0$ and $s^2 = 1$. Define $V_r = r^{-1} \sum_{i=N-r+1}^{N} x_{i:N}$. Prove that

$$(N-r)t^{-1}(N-1)^{-1/2} \le V_r \le \sqrt{(N-r)/r}$$

where $t = \max(r,N-r)$. (Theorems 3.8 and 3.9 are corollaries of this result).

11. Verify the Nair–Thomson lower bound for the range of a sample, equation (3.70), and identify the extremal population. Show that if $i \neq 1$ or $j \neq N$ then $x_{j:N} - x_{i:N}$ can be made arbitrarily small.

12. (Arnold and Groeneveld (1974)). Consider two populations with N_1 and N_2 units. Denote the corresponding elements by x_{ij}, $i = 1,2$; $j = 1,2,\ldots,N_i$. Denote the population means by $\bar{x}_1$ and $\bar{x}_2$. Denote the means of samples of sizes n_1 and n_2 from the two populations by $\bar{x}_1^{(n_1)}$ and $\bar{x}_2^{(n_2)}$ respectively. Verify that

$$\left|(\bar{x}_1^{(n_1)} - \bar{x}_2^{(n_2)}) - (\bar{x}_1 - \bar{x}_2)\right| \le s^*(N_1 + N_2)\sqrt{[n_1^{-1} + n_2^{-1} - 4]},$$

where

$$s^{*2} = \sum_{i=1}^{2} \sum_{j=1}^{N_i} (x_{ij} - \bar{x}_i)^2/(N_1 + N_2).$$

13. (Arnold and Groeneveld (1974)). Let $x_1,\ldots,x_N$ denote a finite population assumed to be symmetric about zero, i.e. $x_{1:N} = -x_{N:N}$, etc. Denote the sample mean based on a sample size n by $\bar{x}^{(n)}$ and denote the population mean and variance by $\bar{x}$ and s^2. Verify that $|\bar{x}^{(n)} - \bar{x}| < s\sqrt{N/(2n)}$. Compare this bound with the general bound provided by Corollary 3.6.

14. (Beesack (1973)). State and prove a version of Theorem 3.11 where f is assumed to be concave on $(0,\infty)$ rather than convex. Use this to verify the following bounds for $x_{j:N}$ corresponding to the case where $f(x) = x^p$, $0 < p < 1$,

$$-2^{-1/p} \le x_{1:N} \le [N - r + r^{1-p}(N-r)^p]^{-1/p},$$

$$-[j + j^p]^{-1/p} \le x_{j:N} \le [N - j + 1 + (N-j+1)^p]^{-1/p}$$

and

$$[r + r^p(N-r)^{1-p}]^{-1/p} \le x_{N:N} \le 2^{-1/p},$$

where r is the number of non-negative x_i's.

15. (Brunk (1959)). An alternative proof of Theorem 3.9 can be based on the following elementary result. If X is a random variable satisfying $0 \le X \le 1$ and $P(X = 1) \ge p$ then $p\,E(X^2) \le [E(X)]^2$. To prove Theorem 3.9, assume without loss of generality that $x_{1:N} = 0$ and $x_{N:N} = 1$ and consider $X = x_{i:N}$ w.p. $1/N$, $i = 1,2,\ldots,N$.

16. (Brunk (1959)). If X is a random variable satisfying $0 \le X \le 1$, it is readily verified that $\mathrm{var}(X) \le 1/4$. We may use this to prove $x_{N:N} - x_{1:N} \ge 2s$. Without loss of generality assume $x_{1:N} = 0$ and $x_{N:N} = 1$ and $X = x_{i:N}$ w.p. $1/N$, $i = 1,2,\ldots,N$.

17. (Nair (1948) and Brunk (1959)). Since $\sum_{i=1}^{N} \sum_{j=i+1}^{N} (x_{j:N} - x_{i:N})$ can be written in the form $\sum_{i=1}^{N} \lambda_i(x_{i:N} - \bar{x})$ for a suitable choice of $\underline{\lambda}$, it can be bounded above using equation (3.57). Verify Nair's result that

$$\sum_{i=1}^{N} \sum_{j=i+1}^{N} (x_{j:n} - x_{i:N}) \le Ns\sqrt{(N^2 - 1)/3}\,.$$

Brunk supplies the following lower bound

$$\sum_{i=1}^{N} \sum_{j=i+1}^{N} (x_{j:N} - x_{i:N}) > Ns\sqrt{N-1}\,.$$

Derive this bound and identify the corresponding extremal population. [Hint: use the following result. If X and Y are i.i.d. random variables with $0 \le X \le 1$, $P(X = 0) \ge p$ and $P(X = 1) \ge p$ ($p < 1/2$), then $E|X - Y|^2 \ge 4p(1-p)\,\mathrm{var}(X)$].

18. Let $x_1',\ldots,x_n'$ denote a random sample from the population $x_1,x_2,\ldots,x_N$. Define $\bar{x}' = \frac{1}{n}\sum_{i=1}^{n} x_i'$ and, as usual, $\bar{x} = \frac{1}{N}\sum_{i=1}^{N} x_i$, $s^2 = \frac{1}{N}\sum_{i=1}^{N} (x_i - \bar{x})^2$. Derive the following bound on the sample mean deviation

$$\frac{1}{n}\sum_{i=1}^{n} |x_i' - \bar{x}'| \le s\sqrt{N/n}\,.$$

(Consider $x_{i1} = \mathrm{sgn}(x_i - \bar{x})$ and $x_{i2} = |x_i - \bar{x}|$ and a suitable choice for $\underline{\lambda}$ in (3.53)).

19. Prove Theorems 3.14 and 3.15.

20. Assume that the population is symmetric about $\bar{x}$. Using the notation of Theorems 3.14 – 3.15 prove

$$|\bar{x}^{(n)} - \bar{x}| \le [x_{N:N} - x_{1:N}]/2, \qquad n \le N/2$$

$\leq [x_{N:N} - x_{1:N}](N-n)/(2n), \quad n > N/2$

and, for positive x_i's,

$|\bar{x}^{(n)} - \bar{x}| \leq \bar{x}.$

21. (Groeneveld (1982)). Assume that $x_i \geq 0$, $\forall$ i. Prove that for $i < j$

$x_{j:N} - x_{i:N} \leq N\bar{x}/(N-j+1).$

22. Suppose $x_i \geq 0$ and $\bar{x} > 0$ then $s(\bar{x}) \leq \bar{x}\sqrt{(N-1)/N}$ $\left[\text{here } [s(\bar{x})]^2 = \frac{1}{N^2}\sum_{i=1}^{N}(x_i - \bar{x})^2\right]$ (Goebel (1974)).

23. (Klamkin (1974)). If the x_i's are not all equal then

$x_{N:N} - \bar{x} \geq s(\bar{x})\sqrt{N/(N-1)}$

where $[s(\bar{x})]$ is as defined in Exercise 22.

24. Verify that the Goebel inequality (Exercise 22) and the Klamkin inequality (Exercise 23) are equivalent.

25. Let $\bar{x}^{(n_1)}$ and $\bar{x}^{(n_2)}$ denote the means of two possibly overlapping samples of sizes n_1 and n_2 from a population of N non-negative units with population mean $\bar{x}$. Prove $|\bar{x}^{(n_1)} - \bar{x}^{(n_2)}|$ $\leq \bar{x}\max\{N/n_1, N/n_2\}$. Interpret this result when $n_1 = n_2 = 1$.

26. Let $\bar{x}^{(n_1)}$ and $\bar{x}^{(n_2)}$ denote the means of two possibly overlapping samples of sizes n_1 and n_2 from a population of N units. Prove

$|\bar{x}^{(n_1)} - \bar{x}^{(n_2)}| \leq s' \max\{N/n_1, N/n_2\}$

where s' is as defined in Corollary 3.12.

27. Guterman (1962) gives a simple proof of (3.67). Denote $(x_{1:N} + x_{N:N})/2$ by m and verify easily that

$$\sum_{i=1}^{N}(x_i - \bar{x})^2 \leq \sum_{i=1}^{N}(x_i - m)^2 \leq N(x_{N:N} - x_{1:N})^2/4.$$

28. Kabe (1980) points out that the bounds in (3.58) can be improved if we are given more information about the x_i's. Suppose we are given

$$s^{*2} = \frac{1}{k}\sum_{i=1}^{k}[x_{i:N} - \bar{x}_{(k)}]^2$$

where

$$\bar{x}_{(k)} = \frac{1}{k} \sum_{i=1}^{k} x_{i:N}.$$

Use the standard two class analysis of variance (total = within + between) to obtain an improved bound on $x_{k:N} - \bar{x}$.

29. Mendenhall (1983, p. 47) states a result which, in our notation, takes the form: Empirical Chebychev Inequality. Among $x_1, x_2, \ldots, x_N$ the number of x_i's which deviate from $\bar{x}$ by more than ks is less than $[N/k^2]$ where $[\cdot]$ denotes integer part.

    (a) Prove the empirical Chebychev inequality (using, if you wish, the usual Chebychev inequality).

    (b) Apply it to the case $k = \sqrt{N}$ to conclude that $\max|x_i - \bar{x}| \le s\sqrt{N}$ (a result slightly weaker than Theorem 3.3). [Kaigh (1980) describes an analogous empirical regression inequality which puts an upper bound on the number of residuals larger than k root mean square residual units.]

30. (David (1981)). Formula (3.91) which is used for example in Theorems 3.18, 3.19 may be replaced by an exact expression for $NE(S^2)$ if the X_i's are assumed to be uncorrelated. Verify that in such a case

$$NE(S^2) = \sum_{i=1}^{N} \left[(\mu_i - \bar{\mu})^2 + \sigma_i^2 \left[\frac{N-1}{N}\right]\right].$$

31. Formula (3.60) might suggest that in general

$$\mu_{1:N} - \bar{\mu} \le -\sigma(N-1)^{-1/2}. \qquad (*)$$

Show that (*) does not always hold. (Hint: consider X_1, X_2 i.i.d. with $P(X_i = 1) = P(X_i = -1) = 1/2$.)

32. Suppose $X_1, X_2, \ldots, X_N$ are jointly distributed with $E(X_i) = \mu \ \forall\, i$ and $\mathrm{var}(X_i) = \sigma^2 \ \forall\, i$. Verify that

    (a) $\mu - \sigma\sqrt{\frac{k_2}{N-k_2}} \le \frac{1}{N-k_1-k_2} \sum_{i=k_1+1}^{k_2} \mu_{i:N} \le \mu + \sigma\sqrt{\frac{k_1}{N-k_1}}.$

    (b) $\mu_{j:N} - \mu_{i:N} \le \sigma\sqrt{\frac{N(N-j+1+i)}{(N-j+1)i}}.$

    (c) $\frac{1}{k} \sum_{i=N-k+1}^{N} \left[\frac{\mu_{i:N} - \mu}{\sigma}\right] \le \sqrt{\frac{N-k}{k}}$ (the left hand expression is the mean selection differential of interest in genetics; see Nagaraja (1981) who also derives a lower bound using the result described in Exercise 10).

33. Suppose $X_1, X_2, \ldots, X_n$ are i.i.d. with common mean 0 and variance 1. By using the Schwarz inequality argument used to derive equation (3.4) verify that

$$\frac{1}{k}\sum_{i=n-k+1}^{n}\mu_{i:n} \le \left[\binom{n}{k}^2 \frac{1}{(2n-1)} \sum_{i,j=n-k+1}^{n} \frac{\binom{n-1}{i-1}\binom{n-1}{j-1}}{\binom{2n-2}{i+j-2}} - 1\right]^{1/2}.$$

This inequality is sharp. Equality obtains for a particular distribution F with bounded support. What is it? (Nagaraja (1981)).

34. Majindar (1962) obtained the result (3.58) in the special case of the median (i.e. i = n+1, N = 2n+1). Use this result and suitable limiting arguments to prove that for any non–degenerate random variable X we have

$$-1 \le [\text{mean }(X) - \text{median }(X)]/\text{s.d. }(X) \le 1$$

a result due to Hotelling and Solomons (1932). (Majindar (1962) and Mallows and Richter (1964) give improved versions of the Hotelling–Solomons result).

35. Consider $X_1, X_2, \ldots, X_n$ jointly distributed with Pareto marginals, i.e.

$$P(X_i > x) = (x/\sigma_i)^{-\alpha}, \quad x > \sigma_i.$$

Verify, using (3.100), that

$$E(X_{n:n}) \le \frac{\alpha}{\alpha - 1}\left[\sum_{i=1}^{n} \sigma_i^{\alpha}\right]^{1/\alpha}.$$

36. (Lai–Robbins (1976) and Gravey (1985)). If $X_1, X_2, \ldots, X_n$ are possibly dependent random variables with $X_i \sim$ exponential (1), $\forall$ i it follows that

$$E(X_{n:n}) \le 1 + \log n.$$

Compare this bound with the known value for $E(X_{n:n})$ in the case of independence.

37. Let $X_1, X_2, \ldots, X_n$ be possibly dependent random variables with common distribution function F. Verify that

$$P(X_{n:n} > x) \le n\bar{F}(x) - \frac{1}{n}\sum_{i \ne j} P(X_i > x, X_j > x)$$

(Galambos (1974/5)). Conclude that if the X_i's are maximally dependent, the events $\{X_i > x\}$ must be disjoint. Discuss these comments in the case where the distributions $\{F_i\}$ are not necessarily the same.

38. Gallot (1966) provides a lower bound for $P(X_{n:n} \ge x)$. We use the notation

$$p_i = P(X_i \ge x), \quad i = 1,2,\ldots,n$$

and

$$P = (p_{ij})_{i,j=1,1}^{n,n}$$

where

$$p_{ij} = P(X_i \ge x, X_j \ge x).$$

Verify that

$$P(X_{n:n} \ge x) \ge \underset{\sim}{p}' P^{-1} \underset{\sim}{p}$$

where $\underset{\sim}{p}' = (p_1, p_2, \ldots, p_n)$. This generally represents an improvement over Whittle's (1959) bound

$$P(X_{n:n} \geq x) \geq \left[\sum_{i=1}^{n} p_i\right]^2 / \left[\sum_{i=1}^{n}\sum_{j=1}^{n} p_{ij}\right].$$

39. (Mallows (1969)). Let $X_1,X_2,\ldots,X_n$ be jointly distributed with $X_i \sim$ uniform (0,1), $\forall$ i. Verify that $E(X_{1:n}) \geq [2n]^{-1}$. (Mallows observes that $\min x_i \geq \sum_{i=1}^{n} [n^{-2} + \min(x_i - n^{-1},0)]$. Compare the proof using this observation with the proof obtainable via Corollary 3.21).

40. Following (3.110) it was observed that if $X_1,X_2,\ldots,X_n$ are possibly dependent uniform (0,1) random variables, then $E(X_{n:n} - X_{1:n}) \leq (n-1)/n$. Verify that this bound is achievable (try a maximally dependent set of X_i's).

41. Suppose $X_1,X_2,\ldots,X_n$ are possibly dependent identically distributed with common distribution F satisfying $\int_0^1 F^{-1}(u)du = 0$ and $\int_0^1 |F^{-1}(u)|^p du = 1$ (here $p > 1$). Verify that

$$E(X_{n:n}) \leq \left[\frac{n(n-1)^{p-1}}{1 + (n-1)^{p-1}}\right]^{1/p},$$

and that this inequality is achievable. [Hint: define $g(u) = nI(u \geq 1-n^{-1})$ and apply the Holder inequality to $\int_0^1 (g-c)F^{-1}du$ then choose c to make the bound as small as possible.]

42. Supply an alternative proof of the result in Exercise 41, using Beesack's inequality (3.76).

43. Suppose $X_1,X_2,\ldots,X_n$ are as in Exercise 41 only now assume, in addition, that F is symmetric. Prove that $E(X_{n:n}) \leq (n/2)^{1/p}$ and that this bound is sharp.

44. Suppose $X_1,X_2,\ldots,X_n$ are i.i.d. symmetric random variables whose common distribution satisfies $\int_0^1 F^{-1}(u)du = 0$ and $\int_0^1 |F^{-1}(u)|^p du = 1$ $(p > 1)$. Verify that

$$E(X_{n:n}) \leq n(\tfrac{1}{2})^{1/p}\, \gamma(n,p)$$

where

$$\gamma(n,p) = \left\{\int_{1/2}^{1} [u^{n-1} - (1-u)^{n-1}]^{p/(p-1)} du\right\}^{(p-1)/p}$$

and that this bound is sharp. (See Arnold (1985) for the corresponding bound when symmetry is not assumed).

45. Repeat Exercises 41 and 44, only this time seek bounds on the expected range.

46. (Extremal cases involving samples from finite populations.) Consider the inequality derived in Exercise 41. Verify that equality obtains if $X_1,X_2,\ldots,X_n$ represent an exhaustive sample drawn without replacement from an urn containing one ball bearing the number

$$[n(n-1)^{p-1}/(1 + (n-1)^{p-1})]^{1/p}$$

and n−1 balls bearing the number $-[n/(1+(n-1)^{p-1})]^{1/p}$. Describe an analogous set of random variables $X_1,X_2,\ldots,X_n$ for which equality obtains in Exercise 43 (note that these vectors $(X_1,\ldots,X_n)$ are maximally dependent).

47. Suppose $X_1,X_2,\ldots,X_n$ are possibly dependent random variables with common uniform (0,1) marginals. For $p > 1$ verify that

$$E(X_{n:n}) \le \left[\frac{n(n-1)^{p-1}}{1+(n-1)^{p-1}}\left[\frac{1}{p+1}\right]\right]^{1/p}.$$

For which p do we get the tightest bound?

48. Suppose $\int_0^1 F^{-1}(u)du = 0$ and $\int_0^1 [F^{-1}(u)]^2 du = 1$.

(a) Verify that for $\alpha \in (0,1)$

$$\alpha F^{-1}(1-\alpha) \le \alpha^{-1/2}$$

[Hint: $\alpha F^{-1}(1-\alpha) \le \int_{1-\alpha}^1 F^{-1}(u)du$, then Schwarz.]

(b) For $t > 0$, verify the one-sided Chebychev inequality

$$F(t) \le (1+t^2)^{-1}.$$

[Hint: Prove the equivalent statement

$$F^{-1}\left[\frac{t^2}{1+t^2}\right] \le t.$$

For it, begin with

$$F^{-1}\left[\frac{t^2}{1+t^2}\right] \le [1+t^2]\int_{t^2/(1+t^2)}^1 F^{-1}(u)du$$

where

$$g(u) = (1+t^2)I(u > t^2/(1+t^2))$$

then use Schwarz].

49. Give an example of a function $\phi:[0,\infty) \to (-\infty,\infty)$ which is star-shaped but not convex.

50. (Barlow and Proschan (1966)). If $F^{-1}G$ is star-shaped ($X_i \sim F$, $Y_i \sim G$) then $E(X_{n-i:n})/E(Y_{n-i:n})$ is increasing in n.

51. (Barlow and Proschan (1966)). If $X_i \ge 0 \sim F$ where F is IFR then $(n-i+1)E(X_{i:n} - X_{i-1:n})$ increases with n for fixed i and decreases with i for fixed n.

52. (Barlow and Proschan (1966)). If $X_i \ge 0 \sim F$ where F is IFRA with mean μ then

$$\mu \sum_{j=1}^{i} (n-j+1)^{-1}\Big/\left[\sum_{j=1}^{i} j^{-1}\right] \le E(X_{i:n}) \le \mu n \sum_{j=1}^{i} (n-j+1)^{-1}$$

for $i = 1,2,\ldots,n-1$.

53. (Nagaraja (1981)). Suppose $X_i \geq 0 \sim F$ having increasing failure rate. It follows that

$$F\left[E\left[\frac{1}{k}\sum_{i=n-k+1}^{n} X_{i:n}\right]\right] \leq 1 - e^{-\sum_{i=k+1}^{n} i^{-1} - 1}$$
$$\leq 1 + \left[\frac{2k+1}{2n+1}\right] e^{-1}.$$

54. (Nagaraja (1981)). Suppose $X_i \sim F$ convex, then

$$F\left[E\left[\frac{1}{k}\sum_{i=n-k+1}^{n} X_{i:n}\right]\right] \leq (2n-k+1)/[2(n+1)].$$

55. (Abdelhamid (1985)). Let $X_1, X_2, \ldots, X_n$ be i.i.d. with distribution F and density f. It follows that

$$\operatorname{var}(X_{n:n}) \leq \frac{1}{2}\int_0^1 (1 - u^n)\,\frac{u}{n}\,[f(F^{-1}(u))]^{-2}du.$$

[Hint: First prove the following theorem of Polya,

$$\int_0^1 g^2(u)du - \left[\int_0^1 g(u)du\right]^2 \leq \frac{1}{2}\int_0^1 u(1-u)[g'(u)]^2du$$

then apply it to $g(u) = F^{-1}_{X_{n:n}}(u)]$. For further discussion see Arnold and Brockett (1988).

56. (Kabir and Rahman (1974)). Equation (3.121) can be improved if, instead of assuming F^{-1} is convex on $(\frac{1}{2}, 1)$, we make the more stringent assumption that $f^*(u) = F^{-1}(u)/(u - \frac{1}{2})$ is convex on $(\frac{1}{2}, 1)$ (an assumption true for normal and t distributions, for example). Verify in this case

$$E(X_{i:n}) \geq \frac{2i - n - 1}{2(n+1)} f^*(c_{i:n})$$

where

$$c_{i:n} = \frac{1}{2} + \frac{2(n+1)}{2i - n - 1}(a_{i:n} - 2b_{i:n})$$

in which

$$a_{i:n} = \frac{i(i+1)}{(n+1)(n+2)} - \frac{i}{n+1} + \frac{1}{4}$$

and

$$b_{i:n} = 2^{-n-2}\binom{n}{i}\sum_{s=i}^{n}\binom{n-i}{s-i}\Big/\binom{s+2}{i}.$$

57. (Patel (1975)). Consider X_i i.i.d. F where for simplicity we assume F has support $[0,\infty)$. There exist a variety of d.f.'s G for which $F^{-1}G$ is convex on the support of G and a variety of d.f.'s H for which $H^{-1}F$ is convex on $[0,\infty)$. Using these we can get a variety of upper and lower bounds for $\sum_{i=1}^{n} \lambda_i E(X_{i:n})$ using (3.116). Illustrate this in the specific case of the mean mid-range $E((X_{1:n}$

$+ X_{n:n})/2)$ from a Weibull distribution (i.e. $F(x) = 1 - e^{-x^\alpha}$ where $\alpha \geq 1$). Select appropriate G's and H's from the following list of d.f.'s: $1 - e^{-x}$, $x > 0$; e^x, $x < 0$; $1 - \frac{1}{x}$, $x > 1$; and x, $0 < x < 1$.

58. (Patel and Read (1975)). Suppose $X_i \geq 0$ are i.i.d. F assumed to be IFR. Define $H(x) = -\log(1 - F(x))$.

(a) Show that for $1 \leq i < j \leq n$

$$E(X_{j:n} | X_{i:n} = x) \leq H^{-1}\left[\sum_{k=i}^{j-1} (n-k)^{-1} + H(x)\right]$$

[Hint: show that for exponential order statistics we have

$$E(Y_{j:n} | Y_{i:n} = y) = \sum_{k=i}^{j-1} (n-k)^{-1} + y$$

and use the implied convexity of F].

(b) Derive analogous bounds for

$E(X_{j:n} | X_{j:n} \leq x < X_{j+1:n})$.

(c) Determine the precise form of the bounds described in (a) and (b) for the special case where F is a Weibull distribution (as described in Exercise 57).

59. (Aven (1985)). Let $X_1,\ldots,X_n$ be possibly dependent. Using the notation of (3.88) prove that for j = 1,2,...,n

$$\mu_{n:n} \leq \bar{\mu} + \sqrt{\frac{n-1}{n}}\left[\tilde{\sigma}_j^2 + \sum_{i=1}^{n} (\mu_i - \bar{\mu})^2\right]^{1/2}$$

and

$$\mu_{n:n} \leq \max_i \mu_i + \sqrt{\frac{n-1}{n}}\,\tilde{\sigma}_j$$

where

$$\tilde{\sigma}_j^2 = \sum_{i=1}^{n} \operatorname{var}(X_i - X_j).$$

60. (Lefevre (1986)). Let $X_1,\ldots,X_n$ be possibly dependent. Using the notation of Theorem 3.18, prove that for j = 1,2,...,n,

$$\left|\sum_{i=1}^{n} \lambda_i(\mu_{i:n} - \bar{\mu})\right| \leq \sqrt{\sum_{i=1}^{n} (\lambda_i - \bar{\lambda})^2}\left[\tilde{\sigma}_j^2 + \sum_{i=1}^{n} (\mu_i - \bar{\mu})^2\right]^{1/2}$$

where $\tilde{\sigma}_j^2$ is as defined in Exercise 59.

61. (David (1986)). Suppose $X_i = Y_i + Z_i$, i = 1,2,...,n are possibly dependent. Prove that

$$E(X_{i:n}) \geq E \max_{j=1,2,\ldots,i} (Y_{j:n} + Z_{i-j+1:n})$$

$$\geq \max_{j=1,2,\ldots,i} [E(Y_{j:n}) + E(Z_{i-j+1:n})].$$

[Hint: the first inequality holds almost surely.]

CHAPTER 4

APPROXIMATIONS TO MOMENTS OF ORDER STATISTICS

4.0. Introduction

In the last chapter we have discussed some universal bounds for the moments of order statistics. In this chapter we shall develop some methods of approximating these moments of order statistics from an arbitrary continuous distribution. First of all, we study the order statistics from a uniform distribution in detail in Section 1 and derive exact and explicit expressions for the single and the product moments. Then, as first noted by K. Pearson and M.V. Pearson (1931) and later worked out in great detail by David and Johnson (1954), these moments of uniform order statistics may be used to develop some simple series approximations for the moments of order statistics from a continuous distribution. (This is to be expected after all in view of the result that the probability integral transformation $U = F(x)$ transforms the order statistic $X_{i:n}$ from any continuous distribution into the order statistic $U_{i:n}$ from a uniform $U(0,1)$ distribution (see Chapter 1).) These series approximations are presented in Section 2. Some similar series expansions have been developed for the first and second order moments of order statistics by Clark and Williams (1958) and these are discussed in Section 3. A different kind of series approximation based on the logistic distribution rather than on the uniform distribution, due to Plackett (1958), is presented in Section 4. In Section 5, we discuss the findings of Saw (1960) who has employed the Darboux form for the remainder in a Taylor series expansion in order to obtain bounds for the remainder term when the expansion for the first single moment $\mu_{i:n}$ $(1 \le i \le n)$ is terminated after an even number of terms in the series. A very interesting and somewhat involved method based on orthogonal inverse expansion which gives approximations as well as bounds for the single and the product moments of order statistics has been developed by Sugiura (1962, 1964). This method of approximation is discussed in great detail in Section 6. Sugiura's method, however, requires that the population have a finite variance. In Section 7, we present a generalization due to Joshi (1969) which provides similar bounds and approximations with less stringent conditions. By noting that these methods do not give satisfactory results for extreme order statistics, Joshi and Balakrishnan (1983) have devised a method by which Sugiura's bounds and approximations may be sharpened considerably, particularly for large sample sizes. This method of improvement for the extreme order statistics is explained in Section 8 and some comparisons of various methods are also made.

4.1. Uniform order statistics and moments

Let $U_1, U_2, \dots, U_n$ be a random sample of size n from the uniform $U(0,1)$ distribution with pdf $f(u) = 1$, $0 \le u \le 1$, and cdf $F(u) = u$, $0 \le u \le 1$. Then the density function of the i'th order statistic $U_{i:n}$ $(1 \le i \le n)$ is given by (eq. (1.7))

$$f_{i:n}(u) = \frac{n!}{(i-1)!(n-i)!} u^{i-1} (1-u)^{n-i}, \quad 0 \le u \le 1$$

from which we get

$$E(U_{i:n}^k) = \frac{n!}{(i-1)!(n-i)!} \int_0^1 u^k u^{i-1} (1-u)^{n-i} du$$

$$= \frac{B(k+i, n-i+1)}{B(i, n-i+1)},$$

where $B(a,b) = \Gamma(a)\,\Gamma(b)/\Gamma(a+b)$, $(a,b > 0)$, is the complete beta function. Upon simplification, we have

$$E(U_{i:n}^k) = \frac{i(i+1)\ldots(i+k-1)}{(n+1)(n+2)\ldots(n+k)}, \quad 1 \le i \le n,\ k \ge 1. \tag{4.1}$$

Setting $k = 1$ in (4.1), we have

$$E(U_{i:n}) = \frac{i}{n+1}, \quad 1 \le i \le n. \tag{4.2}$$

Keeping in mind the probability integral transformation discussed in Chapter 1, we may interpret the result in (4.2) as that the set of n order statistics divide the total area under the curve $y = f(x)$ into $n + 1$ parts, each part with an expected value $1/(n+1)$.

Denoting now $i/(n+1)$ by p_i and $1 - p_i$ by q_i, we have from (4.1) and (4.2) that for $1 \le i \le n$

$$E(U_{i:n}) = p_i \text{ and } Var(U_{i:n}) = p_i q_i/(n+2). \tag{4.3}$$

Similarly, consider the joint density function of $U_{i:n}$ and $U_{j:n}$, $1 \le i < j \le n$, given by (eq. (1.11))

$$f_{i,j:n}(u,v) = \frac{n!}{(i-1)!(j-i-1)!(n-j)!} u^{i-1}(v-u)^{j-i-1}(1-v)^{n-j}, \quad 0 \le u < v \le 1. \tag{4.4}$$

By considering the transformation

$$W_1 = U_{i:n}/U_{j:n} \text{ and } W_2 = U_{j:n}$$

for which the Jacobian is w_2, we get the joint density of W_1 and W_2 from (4.4) as

$$f(w_1,w_2) = \frac{n!}{(i-1)!(j-i-1)!(n-j)!} w_1^{i-1}(1-w_1)^{j-i-1} w_2^{j-1}(1-w_2)^{n-j},$$

$$0 \le w_1 \le 1,\ 0 \le w_2 \le 1. \tag{4.5}$$

It is easy to see from (4.5) that $W_1 = U_{i:n}/U_{j:n}$ and $W_2 = U_{j:n}$ are statistically independent. Moreover, we realize that the marginal distribution of W_1 is Beta $(i,j-i)$ and that of W_2 is Beta $(j,n-j+1)$, which is in accordance with the result in Chapter 1. Using this property of independence of W_1 and W_2, we may now find

$$E(U_{i:n}U_{j:n}) = E(W_1 W_2^2) = E(W_1)\, E(W_2^2)$$

$$= \frac{i(j+1)}{(n+1)(n+2)}$$

and hence

$$Cov(U_{i:n},U_{j:n}) = p_i q_j/(n+2). \tag{4.6}$$

Similarly, starting with the joint density function of $U_{i_r:n}$ $(r = 1,2,..,j,\ 1 \le i_1 < i_2 < \ldots < i_j \le n)$ and then making the transformation $W_r = U_{i_r:n}/U_{i_{r+1}:n}$, $r = 1,2,\ldots,j-1$, and $W_j = U_{i_j:n}$, we shall be able to show that the random variables $W_1, W_2,\ldots, W_j$ are statistically independent and also that the marginal distribution of W_r is Beta $(i_r, i_{r+1}-i_r)$ for $r = 1,2,\ldots,j-1$ and that of W_j is Beta $(i_j, n-i_j+1)$. Making use of this independence, we may find the product moment of a general order as

$$E\left[\prod_{r=1}^{j} U_{i_r:n}^{k_r}\right] = E\left[W_1^{k_1} W_2^{k_1+k_2} \ldots W_j^{k_1+k_2+\ldots+k_j}\right] = \prod_{r=1}^{j} E\left[W_r^{\sum_{\ell=1}^{r} k_\ell}\right]$$

$$= \prod_{r=1}^{j-1} \left\{ \frac{B\left[i_r + \sum_{\ell=1}^{r} k_\ell, i_{r+1} - i_r\right]}{B(i_r, i_{r+1} - i_r)} \right\} \frac{B\left[i_j + \sum_{\ell=1}^{j} k_\ell, n-i_j+1\right]}{B(i_j, n-i_j+1)}$$

$$= \frac{n!}{\left[n + \sum_{\ell=1}^{j} k_\ell\right]!} \prod_{r=1}^{j} \left\{ \frac{\left[i_r - 1 + \sum_{\ell=1}^{r} k_\ell\right]!}{\left[i_r - 1 + \sum_{\ell=1}^{r-1} k_\ell\right]!} \right\}. \tag{4.7}$$

The first four cumulants and cross–cumulants of uniform order statistics have been expressed in terms of p_i and q_i by David and Johnson (1954). These quantities will be used repeatedly in the following sections in order to develop some series approximations for the single and the product moments of order statistics from an arbitrary continuous distribution.

4.2. David and Johnson's approximation

As has been mentioned earlier, the probability integral transformation $u = F(x)$ transforms the order statistic $X_{i:n}$ from a continuous population with pdf $f(x)$ and cdf $F(x)$ into the uniform order statistic $U_{i:n}$. Hence, by inverting the above transformation we have

$$X_{i:n} = F^{-1}(U_{i:n}) = G(U_{i:n}) \tag{4.8}$$

which, upon expanding in a Taylor series about the point $E(U_{i:n}) = p_i$, yields

$$X_{i:n} = G_i + G_i'\,(U_{i:n}-p_i) + \frac{1}{2} G_i''\,(U_{i:n}-p_i)^2 + \frac{1}{6} G_i'''\,(U_{i:n}-p_i)^3 + \frac{1}{24} G_i^{iv}\,(U_{i:n}-p_i)^4 + \ldots; \tag{4.9}$$

G_i denotes $G(p_i)$, G_i' denotes $\frac{d}{du} G(u)\big|_{u=p_i}$, etc. Now taking expectation on both sides of (4.9) and upon using the exact and explicit expressions for the central moments of uniform order statistics, we obtain

$$\mu_{i:n} \simeq G_i + \frac{p_iq_i}{2(n+2)} G_i'' + \frac{p_iq_i}{(n+2)^2} \left[\frac{1}{3}(q_i-p_i)\, G_i''' + \frac{1}{8} p_iq_iG_i^{iv}\right] + \frac{p_iq_i}{(n+2)^3} \left[-\frac{1}{3}(q_i-p_i)G_i''' + \frac{1}{4}\left\{(q_i-p_i)^2 - p_iq_i\right\} G_i^{iv} + \frac{1}{6} q_ip_i(q_i-p_i)\, G_i^{v} + \frac{1}{48} p_i^2 q_i^2 G_i^{vi}\right]. \tag{4.10}$$

Similarly, we may consider the series expansion for $X_{i:n}^2$ obtained from (4.9). Then by taking expectation on both sides and then subtracting the expression for $\mu_{i:n}^2$ obtained from (4.10), we get

$$\mathrm{Var}(X_{i:n}) = \sigma_{i,i:n}$$

$$\simeq \frac{p_iq_i}{(n+2)}(G_i')^2 + \frac{p_iq_i}{(n+2)^2} \left[2(q_i-p_i)\, G_i'\, G_i'' + p_iq_i\left\{G_i'\, G_i''' + \frac{1}{2}(G_i'')^2\right\}\right]$$

$$+ \frac{p_iq_i}{(n+2)^3} \left[-2(q_i-p_i)\, G_i'G_i'' + \left\{(q_i-p_i)^2 - p_iq_i\right\}\right.$$

$$\left\{2G_i' G_i''' + \frac{3}{2}(G_i'')^2\right\} + p_iq_i(q_i-p_i)(\tfrac{5}{3}G_i'G_i^{iv} + 3G_i''G_i''')$$
$$+ \tfrac{1}{4}p_i^2q_i^2\left\{G_i'G_i^{v} + 2G_i''G_i^{iv} + \tfrac{5}{3}(G_i''')^2\right\}\Big]. \tag{4.11}$$

By considering the series expansion for $X_{i:n}X_{j:n}$ obtained from (4.9) and proceeding on similar lines, we derive

$$\begin{aligned}
\mathrm{Cov}(X_{i:n}, X_{j:n}) &= \sigma_{i,j:n} \\
&\simeq \frac{p_iq_j}{(n+2)} G_i'G_j' + \frac{p_iq_j}{(n+2)^2}\Big[(q_i-p_i)G_i''G_j' + (q_j-p_j)G_i'G_j'' \\
&+ \tfrac{1}{2}p_iq_iG_i'''G_j' + \tfrac{1}{2}p_jq_jG_i'G_j''' + \tfrac{1}{2}p_iq_jG_i''G_j''\Big] \\
&+ \frac{p_iq_j}{(n+2)^3}\Big[-(q_i-p_i)G_i''G_j' - (q_j-p_j)G_i'G_j'' + \{(q_i-p_i)^2 - p_iq_i\}G_i'''G_j' \\
&+ \{(q_j-p_j)^2 - p_jq_j\}G_i'G_j''' + \{\tfrac{3}{2}(q_i-p_i)(q_j-p_j) + \tfrac{1}{2}p_jq_i - 2p_iq_j\}G_i''G_j'' \\
&+ \tfrac{5}{6}p_iq_i(q_i-p_i)G_i^{iv}G_j' + \tfrac{5}{6}p_jq_j(q_j-p_j)G_i'G_j^{iv} + \{p_iq_j(q_i-p_i) \\
&+ \tfrac{1}{2}p_iq_i(q_j-p_j)\}G_i'''G_j'' + \{p_iq_j(q_j-p_j) + \tfrac{1}{2}p_jq_j(q_i-p_i)\}G_i''G_j''' \\
&+ \tfrac{1}{8}p_i^2q_i^2G_i^{v}G_j' + \tfrac{1}{8}p_j^2q_j^2G_i'G_j^{v} + \tfrac{1}{4}p_i^2q_iq_jG_i^{iv}G_j'' \\
&+ \tfrac{1}{4}p_ip_jq_j^2G_i''G_j^{iv} + \tfrac{1}{12}(2p_i^2q_j^2 + 3p_iq_ip_jq_j)G_i'''G_j'''\Big]
\end{aligned} \tag{4.12}$$

David and Johnson (1954) have worked out similar series approximations for the first four cumulants and cross–cumulants of order statistics. Realize that they have given these approximations in inverse powers of n + 2 (up to order 3) as they found it then convenient to combine terms of higher order. It is also important to note that

$$G_i' = \frac{d}{du}G(u)\Big|_{u=p_i} = \frac{1}{f(F^{-1}(u))}\Big|_{u=p_i} = \frac{1}{f(G_i)} \tag{4.13}$$

which is just the reciprocal of the probability density function of X evaluated at G_i. This form of G_i' in (4.13) allows us to write down the higher order derivatives of G_i without great difficulty in most cases. For example, for the standard normal distribution, making use of the property that $f'(x) = -x\,f(x)$ we have

$$G_i' = \frac{1}{f(G_i)},\quad G_i'' = \frac{G_i}{\{f(G_i)\}^2},\quad G_i''' = \frac{1+2G_i^2}{\{f(G_i)\}^3},$$
$$G_i^{iv} = \frac{G_i(7+6G_i^2)}{\{f(G_i)\}^4},\quad G_i^{v} = \frac{7+46G_i^2+24G_i^4}{\{f(G_i)\}^5},$$
$$G_i^{vi} = \frac{G_i(127+326G_i^2+120G_i^4)}{\{f(G_i)\}^6},\ \text{etc.}$$

For the case of the logistic distribution with

$$f(x) = \frac{e^{-x}}{(1+e^{-x})^2} \text{ and } F(x) = \frac{1}{1+e^{-x}},\qquad -\infty < x < \infty, \tag{4.14}$$

however, G_i and its derivatives may be worked out rather easily. This is to be expected after all as the logistic distribution admits an exact and explicit form for the inverse cumulative distribution function. For example, we have in this case

$G_i = \ln(p_i) - \ln(q_i), \quad G_i' = \frac{1}{p_i} + \frac{1}{q_i},$

$G_i'' = -\frac{1}{p_i^2} + \frac{1}{q_i^2}, \quad G_i''' = 2\left\{\frac{1}{p_i^3} + \frac{1}{q_i^3}\right\},$

$G_i^{iv} = 6\left\{-\frac{1}{p_i^4} + \frac{1}{q_i^4}\right\}, \quad G_i^{v} = 24\left\{\frac{1}{p_i^5} + \frac{1}{q_i^5}\right\},$

$G_i^{vi} = 120\left\{-\frac{1}{p_i^6} + \frac{1}{q_i^6}\right\}$, etc.

As illustrated by David and Johnson (1954) and also by various other authors, this simple and straight forward approximation procedure works well in most cases. However, the method does not usually provide satisfactory results for extreme order statistics. In the case of extreme order statistics, the convergence of this approximation to the true value may be very slow and even nonexistent in some cases. One may refer to Section 8 for more details on this issue.

4.3. Clark and Williams' approximation

Clark and Williams (1958) have developed some series approximations which are quite similar to those of David and Johnson (1954). Approximations given by Clark and Williams (1958) make use of the exact expressions of the central moments of uniform order statistics where the i'th central moment is of order $\{(n+2)(n+3)...(n+i)\}^{-1}$; David and Johnson's (1954) approximations, on the other hand, are based on the approximate expressions of the central moments of uniform order statistics which are in inverse powers of n+2, as mentioned earlier in Section 2.

Thus, by starting with the series expansion of $X_{i:n}$ given in (4.9) and then by using equation (4.13) and the exact expressions of the central moments of uniform order statistics, we obtain

$$E(X_{i:n}) = \mu_{i:n}$$
$$= G_i - \frac{f_i'}{2f_i^3}\frac{p_iq_i}{(n+2)} + \left\{\frac{3(f_i')^2 - f_if_i''}{3f_i^5}\right\}\frac{p_iq_i(q_i - p_i)}{(n+2)(n+3)}$$
$$+ \left\{\frac{10f_if_i'f_i'' - f_i^2f_i''' - 15(f_i')^3}{8f_i^7}\right\}\frac{np_i^2q_i^2 + p_iq_i(2-5p_iq_i)}{(n+2)(n+3)(n+4)} + \ldots, \tag{4.15}$$

where $G_i = G(p_i)$, $f_i = f(G_i)$, $f_i' = f'(G_i)$, etc. Proceeding similarly, we also obtain

$$\mathrm{Var}(X_{i:n}) = \frac{1}{f_i^2}\frac{p_iq_i}{(n+2)} - \frac{f_i'}{f_i^4}\frac{2p_iq_i(q_i - p_i)}{(n+2)(n+3)} + \left[\frac{15(f_i')^2}{4f_i^6} - \frac{f_i''}{f_i^5}\right]\frac{np_i^2q_i^2 + p_iq_i(2-5p_iq_i)}{(n+2)(n+3)(n+4)}$$
$$- \frac{(f_i')^2}{4f_i^6}\frac{p_i^2q_i^2}{(n+2)^2} + \ldots, \tag{4.16}$$

$$E(X_{i:n}-\mu_{i:n})^3 = \frac{1}{f_i^3}\frac{2p_iq_i(q_i-p_i)}{(n+2)(n+3)} - \frac{9f_i'}{2f_i^5}\frac{np_i^2q_i^2 + p_iq_i(2-5p_iq_i)}{(n+2)(n+3)(n+4)}$$

$$+\frac{3f_i'}{2f_i^5}\frac{p_i^2q_i^2}{(n+2)^2}+\ldots, \tag{4.17}$$

and

$$E(X_{i:n}-\mu_{i:n})^4=\frac{3}{f_i^4}\frac{np_i^2q_i^2+p_iq_i(2-5p_iq_i)}{(n+2)(n+3)(n+4)}+\ldots. \tag{4.18}$$

From the above results we easily see the well known fact that if both i and n increase with i/n remaining fixed, the asymptotic distribution of $X_{i:n}$ has the mean and variance as G_i and p_iq_i/nf_i^2, respectively. Furthermore, from equations (4.16) – (4.18) we also have

$$\frac{E(X_{i:n}-\mu_{i:n})^3}{\{Var(X_{i:n})\}^{3/2}}=\frac{1}{\sqrt{n}}\left[\frac{2(q_i-p_i)}{\sqrt{p_iq_i}}-\frac{3\sqrt{p_iq_i}\;f_i'}{f_i^2}\right]+\ldots \tag{4.19}$$

and

$$\frac{E(X_{i:n}-\mu_{i:n})^4}{\{Var(X_{i:n})\}^2}=3\left\{1-\frac{5n+12}{(n+3)(n+4)}\right\}+\ldots. \tag{4.20}$$

Realize that the known result that for large n the distribution of $X_{i:n}$ is approximately normal is apparent from equations (4.19) and (4.20).

Starting similarly with the series expansion of $X_{i:n}X_{j:n}$ and then using the exact expressions of the central product moments of uniform order statistics, Clark and Williams (1958) have also derived approximations for the covariance and the correlation coefficient between the order statistics $X_{i:n}$ and $X_{j:n}$ as

$$\begin{aligned}
Cov(X_{i:n},X_{j:n}) &= \frac{1}{f_if_j}\frac{p_iq_j}{(n+2)}-\frac{f_i'}{f_i^3f_j}\frac{p_iq_j(q_i-p_i)}{(n+2)(n+3)}-\frac{f_j'}{f_if_j^3}\frac{p_iq_j(q_j-p_j)}{(n+2)(n+3)} \\
&+\left\{\frac{3(f_i')^2-f_if_i''}{2f_i^5f_j}\right\}\frac{np_i^2q_iq_j+p_iq_j(2-5p_iq_i)}{(n+2)(n+3)(n+4)} \\
&+\left\{\frac{3(f_j')^2-f_jf_j''}{2f_if_j^5}\right\}\frac{np_ip_jq_j^2+p_iq_j(2-5p_jq_j)}{(n+2)(n+3)(n+4)} \\
&+\frac{f_i'f_j'}{4f_i^3f_j^3}\frac{np_iq_j(q_i-q_j+3p_iq_j)+p_iq_j\{1+5(p_i+q_j)-15p_iq_j\}}{(n+2)(n+3)(n+4)} \\
&-\left[\frac{f_i'}{2f_i^3}\frac{p_iq_i}{(n+2)}+\left\{\frac{3(f_i')^2-f_i''}{3f_i^5}\right\}\frac{p_iq_i(q_i-p_i)}{(n+2)(n+3)}\right. \\
&\left.+\left\{\frac{10f_if_i'f_i''-f_i^2f_i'''-15(f_i')^3}{8f_i^7}\right\}\frac{np_i^2q_i^2+p_iq_i(2-5p_iq_i)}{(n+2)(n+3)(n+4)}\right] \\
&-\left[\frac{f_j'}{2f_j^3}\frac{p_jq_j}{(n+2)}+\left\{\frac{3(f_j')^2-f_j''}{3f_j^5}\right\}\frac{p_jq_j(q_j-p_j)}{(n+2)(n+3)}\right. \\
&\left.+\left\{\frac{10f_jf_j'f_j''-f_j^2f_j'''-15(f_j')^3}{8f_j^7}\right\}\frac{np_j^2q_j^2+p_jq_j(2-5p_jq_j)}{(n+2)(n+3)(n+4)}\right]+\ldots
\end{aligned} \tag{4.21}$$

and

$$\rho(X_{i:n},X_{j:n}) = \left[\frac{p_iq_j}{q_ip_j}\right]^{1/2}\left[1 - \frac{1}{4(n+2)}\left\{\frac{(f_i')^2}{f_i^4}p_iq_i - \frac{2f_i'f_j'}{f_i^2f_j^2}p_iq_j + \frac{(f_j')^2}{f_j^4}p_jq_j\right\}\right] + \ldots . \tag{4.22}$$

Formula in (4.22) has also been derived by David and Johnson (1954) from equations (4.11) and (4.12). A comparison of the respective formulae in Sections 2 and 3 reveals that the approximations provided by David ad Johnson (1954) are lot easier to use in practice.

4.4. Plackett's approximation

The methods of approximation discussed in Sections 2 and 3 have both been developed by applying the probability integral transformation and then using the known moments of order statistics from the uniform distribution. Plackett (1958), instead, has proposed a method based on the logit transformation which transforms a continuous order statistic $X_{i:n}$ into the order statistic $L_{i:n}$ from the standard logistic distribution and then develops some approximations by using the explicit expressions of the moments of logistic order statistics.

To illustrate the method of approximation due to Plackett (1958), let us denote $L_{i:n}$ ($1 \le i \le n$) for the i'th order statistic in a sample of size n from a standard logistic distribution with pdf and cdf as given in (4.14). Then the moment generating function of $L_{i:n}$ is given by

$$\begin{aligned} E\left[e^{tL_{i:n}}\right] &= \frac{n!}{(i-1)!(n-i)!}\int_0^1 \left[\frac{u}{1-u}\right]^t u^{i-1}(1-u)^{n-i}\,du \\ &= \frac{n!}{(i-1)!(n-i)!}\,\frac{\Gamma(i+t)\,\Gamma(n-i-t+1)}{\Gamma(n+1)} \\ &= \Gamma(i+t)\,\Gamma(n-i-t+1)/\Gamma(i)\,\Gamma(n-i+1). \end{aligned} \tag{4.23}$$

From (4.23), by taking logarithms, differentiating with respect to t and then setting $t = 0$, we obtain the cumulants of $L_{i:n}$ as

$$\begin{aligned} \kappa_{i:n}^{(j)} &= \frac{d^j}{dt^j}\ln\Gamma(i+t) + \frac{d^j}{dt^j}\ln\Gamma(n-i+1-t)\Big|_{t=0} \\ &= \Psi^{(j-1)}(i) + (-1)^j\,\Psi^{(j-1)}(n-i+1), \end{aligned} \tag{4.24}$$

where $\Psi^{(0)}(z) = \Psi(z) = \frac{d}{dz}\ln\Gamma(z) = \Gamma'(z)/\Gamma(z)$ is the psi or digamma function, and $\Psi^{(j-1)}(z) = \frac{d^{j-1}}{dz^{j-1}}\Psi(z)$ are the derivatives of the psi function which are referred to as polygamma functions. These functions have been tabulated quite extensively by Davis (1935) and Abramowitz and Stegun (1965). For $i > (n+1)/2$, for example, we derive from (4.24) that

$$\kappa_{i:n}^{(1)} = \sum_{k=n-i+1}^{i-1} 1/k, \tag{4.25}$$

$$\kappa_{i:n}^{(2)} = \frac{\pi^2}{3} - \sum_{k=1}^{i-1} 1/k^2 - \sum_{k=1}^{n-i} 1/k^2, \tag{4.26}$$

$$\kappa_{i:n}^{(3)} = 2 \sum_{k=n-i+1}^{i-1} 1/k^3 \tag{4.27}$$

and

$$\kappa_{i:n}^{(4)} = \frac{2\pi^4}{15} - 6 \sum_{k=1}^{i-1} 1/k^4 - 6 \sum_{k=1}^{n-i} 1/k^4. \tag{4.28}$$

As has been mentioned earler, the logit transformation $L = \ln[F(x)/\{1-F(x)\}]$ transforms the order statistic $X_{i:n}$ from a continuous population with pdf f(x) and cdf F(x) into the logistic order statistic $L_{i:n}$. Then by considering $X_{i:n}$ as a function of $L_{i:n}$ and expanding in a Taylor series about the point $E(L_{i:n}) = \kappa_{i:n}^{(1)}$, we have

$$X_{i:n} = x^{(0)} + x^{(1)} \left[L_{i:n} - \kappa_{i:n}^{(1)}\right] + \frac{1}{2} x^{(2)} \left[L_{i:n} - \kappa_{i:n}^{(1)}\right]^2 + \frac{1}{6} x^{(3)} \left[L_{i:n} - \kappa_{i:n}^{(1)}\right]^3 + \frac{1}{24} x^{(4)} \left[L_{i:n} - \kappa_{i:n}^{(1)}\right]^4 + \dots, \tag{4.29}$$

where $x^{(j)}$ is the value of the j'th derivative of x with respect to L at $L = \kappa_{i:n}^{(1)}$. Now taking expectation on both sides of (4.29) and upon using the exact and explicit expressions for the cumulants of logistic order statistics given in (4.24) – (4.28), we obtain

$$\mu_{i:n} \simeq x^{(0)} + \frac{1}{2} x^{(2)} \kappa_{i:n}^{(2)} + \frac{1}{6} x^{(3)} \kappa_{i:n}^{(3)} + \frac{1}{24} x^{(4)} \left\{\kappa_{i:n}^{(4)} + 3\left[\kappa_{i:n}^{(2)}\right]^2\right\}. \tag{4.30}$$

The coefficients in the above approximation are easy to obtain as in the case of David and Johnson's (1954) approximation. For example, for the standard normal distribution with pdf f(x) and cdf F(x), we have

$$x^{(0)} = F^{-1}\left\{e^{\kappa_{i:n}^{(1)}}/(1+e^{\kappa_{i:n}^{(1)}})\right\},$$

$$x^{(1)} = F(1-F)/f,$$

$$x^{(2)} = x^{(1)}\{xx^{(1)} - (2F-1)\},$$

$$x^{(3)} = (x^{(1)})^3 + 2xx^{(1)}x^{(2)} + x^{(2)}(1-2F) - 2x^{(1)}F(1-F)$$

and

$$x^{(4)} = 5(x^{(1)})^2 x^{(2)} + x^{(3)}\{2xx^{(1)} - (2F-1)\} + 2x^{(2)}\{xx^{(2)} - 2F(1-F)\} + 2x^{(1)} (2F-1)F(1-F);$$

these derivatives are all bounded. Suppose we include the first j–1 terms in the series expansion for $\mu_{i:n}$ given in equation (4.30), then the absolute value of the remainder after j–1 terms is at most $\frac{1}{j!} \max \left|x^{(j)}\right| E\left|L_{i:n} - \kappa_{i:n}^{(1)}\right|^j$. Since $E\left|L_{i:n} - \kappa_{i:n}^{(1)}\right|^{2j}$ is known and also that

$$\left\{E\left|L_{i:n} - \kappa_{i:n}^{(1)}\right|^{2j-1}\right\}^{\frac{1}{2j-1}} \le \left\{E\left|L_{i:n} - \kappa_{i:n}^{(1)}\right|^{2j}\right\}^{\frac{1}{2j}},$$

we realize that we will be able to present bounds to $\mu_{i:n}$ for all values of j.

4.5. Saw's error analysis

We shall consider here David and Johnson's (1954) series approximation for the mean of normal order statistic discussed in Section 2 and illustrate Saw's (1960) method of deriving bounds for the error which results in using only a finite number of terms in the series.

From (4.8), we may write

$$\mu_{i:n} = E\left[X_{i:n}\right] = \sum_{t=0}^{2m} \frac{1}{t!} G_i^{(t)} E(U_{i:n}-p_i)^t + R_{2m}, \tag{4.31}$$

where $p_i = E(U_{i:n}) = i/(n+1)$, $G(u) = F^{-1}(u)$, $G_i = G(p_i)$ and $G_i^{(t)} = \frac{d^t}{du^t} G(u)|_{u=p_i}$; R_{2m}, the remainder after 2m terms, is the difference between the true value of $\mu_{i:n}$ and the sum of the series to 2m terms. Let us consider the Taylor series expansion for the function $x = F^{-1}(u) = G(u)$ about the point $u = p_i$ given by

$$x = G(u) = \sum_{t=0}^{2m} \frac{1}{t!} G_i^{(t)} (u-p_i)^t + R'_{2m}. \tag{4.32}$$

Using the Darboux form for the remainder R'_{2m} in (4.32), we may write

$$R'_{2m} = \frac{1}{(2m)!} \int_0^1 (1-\xi)^{2m} G^{(2m+1)}(p_i + \xi(u-p_i))\, d\xi. \tag{4.33}$$

Now multiplying both sides of (4.32) by the function $u^{i-1}(1-u)^{n-i}/B(i,n-i+1)$, integrating over u in the interval (0,1), and comparing the resulting equation with (4.31), we get

$$R_{2m} = \int_0^1 \frac{1}{B(i,n-i+1)} u^{i-1} (1-u)^{n-i} R'_{2m}(u)\, du;$$

combining this with equation (4.33), we have

$$R_{2m} = \frac{1}{(2m)!} \int_0^1 \int_0^1 (1-\xi)^{2m} G^{(2m+1)}\left[p_i+\xi(u-p_i)\right] \frac{1}{B(i,n-i+1)} u^{i-1}(1-u)^{n-i}\, du\, d\xi. \tag{4.34}$$

As noted in Section 2, we see that for $j \geq 1$

$$\frac{d^j}{du^j} x(u) = G^{(j)}(u) = H_j(x)/(f(x))^j,$$

where $H_j(x)$ is a polynomial of degree j−1 in x which is an even or odd function depending on whether j is odd or even. We have, for example,

$$H_1(x) = 1,\ H_2(x) = x,\ H_3(x) = 1+2x^2,$$

$$H_4(x) = 7x + 6x^3,\ H_5(x) = 7 + 46x^2 + 24x^4,$$

$$H_6(x) = 127x + 326x^3 + 120x^5,$$

etc., and it is easily verified that these polynomials satisfy the recurrence relation

$$H_{j+1}(x) = j\, x\, H_j(x) + H'_j(x), \tag{4.35}$$

where $H'_j(x) = \frac{d}{dx} H_j(x)$. We also note that

$$G^{(j)}(u) = (-1)^{j-1} G^{(j)}(1-u). \tag{4.36}$$

Denoting now $G_j^*(u)$ for $u^j G^{(j)}(u)$, we see immediately that $G_j^*(0) = 0$ and $G_j^*(u) \to \infty$ as $u \to 1$.

Moreover, for $\frac{1}{2} < u < 1$, we note that $G^*_{2m+1}(u) = u^{2m+1}\, G^{(2m+1)}(u)$ is the product of two increasing functions and, hence, is an increasing function of u. Next, for $0 < u < \frac{1}{2}$ (that is, $x < 0$), we consider

$$\begin{aligned}\frac{d}{du} G^*_{2m+1}(u) &= (2m+1)\, u^{2m}\, G^{(2m+1)}(u) + u^{2m+1}\, G^{(2m+2)}(u)\\ &= (2m+1)\, u^{2m} \frac{H_{2m+1}(x)}{(f(x))^{2m+1}} + u^{2m+1} \frac{H_{2m+2}(x)}{(f(x))^{2m+2}}\\ &= \frac{u^{2m}}{(f(x))^{2m+2}} H_{2m+2}(x) \left\{(2m+1)\, f(x) \frac{H_{2m+1}(x)}{H_{2m+2}(x)} + u\right\}\\ &= K_1(u)\, K_2(u), \end{aligned} \tag{4.37}$$

where

$$K_1(u) = \frac{u^{2m}}{(f(x))^{2m+2}} H_{2m+2}(x) \quad \text{and} \quad K_2(u) = (2m+1) f(x) \frac{H_{2m+1}(x)}{H_{2m+2}(x)} + u.$$

Now for $0 < u < \frac{1}{2}$, we have $x < 0$ and, consequently, $K_1(u) < 0$. Noting that $K_2(0) = 0$, we consider

$$\begin{aligned}\frac{d\, K_2(u)}{dx} &= \frac{-f(x)}{H^2_{2m+2}(x)} \Big[- H^2_{2m+2}(x) - (2m+1)\, H'_{2m+1}(x)\, H_{2m+2}(x)\\ &\qquad + (2m+1)\, x\, H_{2m+1}(x)\, H_{2m+2}(x) + (2m+1)\, H_{2m+1}(x)\, H'_{2m+2}(x)\Big]\\ &= -f(x)\, A_{2m+1}(x)/H^2_{2m+2}(x) \end{aligned} \tag{4.38}$$

upon using the relation in (4.35), where

$$A_{2m+1}(x) = (2m+1)\, H_{2m+1}(x)\, H'_{2m+2}(x) - (2m+2)\, H'_{2m+1}(x)\, H_{2m+2}(x)$$

or, equivalently,

$$A_{2m+1}(x) = (2m+1)\, H_{2m+1}(x)\, H_{2m+3}(x) - (2m+2)\, H^2_{2m+2}(x). \tag{4.39}$$

From (4.39), for example, we get

$$A_1(x) = 1,$$

$$A_3(x) = 21 - 16\, x^2 + 12\, x^4,$$

$$A_5(x) = 4445 - 6664\, x^2 + 8076\, x^4 + 1344\, x^6 + 2880\, x^8,$$

$$\begin{aligned}A_7(x) &= 3{,}884{,}041 - 8{,}666{,}072\, x^2 + 14{,}468{,}040\, x^4\\ &\quad + 6{,}808{,}896\, x^6 + 24{,}184{,}656\, x^8 + 11{,}093{,}760\, x^{10} + 3{,}628{,}800\, x^{12},\end{aligned}$$

$$\begin{aligned}A_9(x) &= 9{,}580{,}522{,}329 - 28{,}374{,}712{,}624\, x^2 + 60{,}246{,}981{,}384\, x^4\\ &\quad + 54{,}671{,}037{,}120\, x^6 + 292{,}158{,}113{,}616\, x^8 + 337{,}984{,}717{,}824\, x^{10}\\ &\quad + 259{,}516{,}161{,}792\, x^{12} + 88{,}044{,}104{,}640\, x^{14} + 14{,}631{,}321{,}600\, x^{16}.\end{aligned}$$

We note that $A_{2m+1}(x)$ is positive definite in all these cases since the first three terms (that is, the terms involving x^0, x^2 and x^4) form a positive definite quadratic in x^2. Based on (4.37) and (4.38), therefore, we conclude that $G^*_{2m+1}(u)$ is an increasing function in u in the interval (0,1) (at least for the cases m = 0,1,2,3,4).

For the standard normal distribution, since $\mu_{i:n} = -\mu_{n-i+1:n}$ let us consider the cases when $i > (n+1)/2$, that is, $p_i > 1/2$. Making a substitution $v = p_i + \xi(u-p_i)$ in equation (4.34), we get

$$R_{2m} = \frac{1}{(2m)!}\left[-\int_0^{p_i}\int_0^{v} (u-v)^{2m}\, G^{(2m+1)}(v)\, \frac{1}{B(i,n-i+1)}\, u^{i-1}(1-u)^{n-i}\, du\, dv \right.$$
$$\left. + \int_0^{p_i}\int_v^{1} (u-v)^{2m}\, G^{(2m+1)}(v)\, \frac{1}{B(i,n-i+1)}\, u^{i-1}(1-u)^{n-i}\, du\, dv\right].$$

Replacing v by 1–v, u by 1–u and then using the symmetrical property of $G^{(2m+1)}(v)$ given in (4.36) in the second integral, we obtain

$$R_{2m} = \frac{1}{(2m)!}\iint\limits_{0<u<v<q_i} (v-u)^{2m}\, G^{(2m+1)}(v)\, \frac{1}{B(i,n-i+1)}\left\{u^{n-i}(1-u)^{i-1} - u^{i-1}(1-u)^{n-i}\right\} dv\, du$$
$$- \frac{1}{(2m)!}\int_{q_i}^{p_i}\int_0^{v} (v-u)^{2m}\, G^{(2m+1)}(v)\, \frac{1}{B(i,n-i+1)}\, u^{i-1}(1-u)^{n-i}\, du\, dv$$
$$= J_1 - J_2 \text{ (say)}; \tag{4.40}$$

note that J_1 and J_2 are both positive since the integrands are everywhere positive. Now by considering

$$J_1 = \frac{1}{(2m)!}\iint\limits_{0<u<v<q_i} (1-\tfrac{u}{v})^{2m}\, \frac{1}{v}\, G^*_{2m+1}(v)\, \frac{1}{B(i,n-i+1)}\left\{u^{n-i}(1-u)^{i-1} - u^{i-1}(1-u)^{n-i}\right\} dv\, du$$

and using the property that $G^*_{2m+1}(v)$ is an increasing function in v, we have

$$J_1 < \frac{G^*_{2m+1}(q_i)}{(2m)!}\iint\limits_{0<u<v<q_i} (1-\tfrac{u}{v})^{2m}\, \frac{1}{v}\, \frac{1}{B(i,n-i+1)}\left\{u^{n-i}(1-u)^{i-1} - u^{i-1}(1-u)^{n-i}\right\} dv\, du.$$

Since

$$\int_u^{q_i}(1-\tfrac{u}{v})^{2m}\frac{dv}{v} < \int_u^{q_i}(1-\tfrac{u}{v})^{2m}\frac{q_i}{v^2}\, dv = \frac{(q_i - u)^{2m+1}}{(2m+1)\, q_i^{2m}\, u},$$

we have

$$J_1 < \frac{q_i\, G_i^{(2m+1)}}{(2m+1)!}\int_0^{q_i}(q_i-u)^{2m+1}\,\frac{1}{u}\,\frac{1}{B(i,n-i+1)}\left\{u^{n-i}(1-u)^{i-1} - u^{i-1}(1-u)^{n-i}\right\} du. \tag{4.41}$$

By considering J_1 once again from (4.40) and using the property that $G^{(2m+1)}(v)$ is a decreasing function of v in $(0, q_i < \frac{1}{2})$, we have

$$J_1 > \frac{G_i^{(2m+1)}}{(2m+1)!}\iint\limits_{0<u<v<q_i} (v-u)^{m+1}\, \frac{1}{B(i,n-i+1)}\left\{u^{n-i}(1-u)^{i-1} - u^{i-1}(1-u)^{n-i}\right\} du\, dv$$

which, when performed the integration over v, yields

$$J_1 > \frac{G_i^{(2m+1)}}{(2m+1)!}\int_0^{q_i}(q_i-u)^{2m+1}\,\frac{1}{B(i,n-i+1)}\left\{u^{n-i}(1-u)^{i-1} - u^{i-1}(1-u)^{n-i}\right\} du. \tag{4.42}$$

By considering J_2 from (4.40) and using the property that $G^{(2m+1)}(v)$ is an increasing function of v^2, we have

$$J_2 < \frac{G_i^{(2m+1)}}{(2m)!}\int_{q_i}^{p_i}\int_0^{v}(v-u)^{2m}\,\frac{1}{B(i,n-i+1)}\, u^{i-1}(1-u)^{n-i}\, du\, dv$$

and

$$J_2 > \frac{G^{(2m+1)}(\frac{1}{2})}{(2m)!} \int_{q_i}^{p_i} \int_0^v (v-u)^{2m} \frac{1}{B(i,n-i+1)} u^{i-1}(1-u)^{n-i}\, du\, dv.$$

Writing both the double integrals as

$$\int_{q_i}^{p_i} \int_0^v du\, dv \equiv \int_{q_i}^{p_i} dv \int_0^{q_i} du + \int_{q_i}^{p_i} \int_u^{p_i} dv\, du$$

and integrating out over v, we obtain after simplification

$$J_2 < \frac{-G_i^{(2m+1)}}{(2m+1)!} \Big[E(U_{i:n} - p_i)^{2m+1} + \int_0^{q_i} (u - q_i)^{2m+1} \frac{1}{B(i,n-i+1)} \left\{u^{n-i}(1-u)^{i-1} - u^{i-1}(1-u)^{n-i}\right\} du\Big] \tag{4.43}$$

and

$$J_2 > \frac{-G^{(2m+1)}(\frac{1}{2})}{(2m+1)!} \Big[E(U_{i:n}-p_i)^{2m+1} + \int_0^{q_i} (u-q_i)^{2m+1} \frac{1}{B(i,n-i+1)} \left\{u^{n-i} (1-u)^{i-1} - u^{i-1}(1-u)^{n-i}\right\} du\Big]. \tag{4.44}$$

Combining equations (4.42) and (4.43), we obtain from (4.40) that

$$R_{2m} > \frac{G_i^{(2m+1)}}{(2m+1)!} E(U_{i:n} - p_i)^{2m+1}. \tag{4.45}$$

Similarly, combining equations (4.41) and (4.44) we derive an upper bound for R_{2m} from (4.40) as

$$R_{2m} < \frac{q_i G_i^{(2m+1)}}{(2m+1)!} \left\{V_{2m+1}(i,n-i+1;q_i) - V_{2m+1}(n-i+1,i;q_i)\right\}$$
$$- \frac{G^{(2m+1)}(\frac{1}{2})}{(2m+1)!} \left\{W_{2m+1}(i,n-i+1;q_i) - W_{2m+1}(n-i+1,i;q_i) - E(U_{i:n} - p_i)^{2m+1}\right\}, \tag{4.46}$$

where

$$W_c(a,b;v) = \int_0^v (u-v)^c \frac{1}{B(a,b)} u^{a-1} (1-u)^{b-1}\, du$$

and

$$V_c(a,b;v) = \int_0^v \frac{1}{u} (u-v)^c \frac{1}{B(a,b)} u^{a-1} (1-u)^{b-1} du.$$

For computational ease, we give the following recurrence relations satisfied by the above quantities involved in computing the bounds derived in (4.45) and (4.46):

$$\frac{n+c+2}{c} E(U_{i:n} - p_i)^{c+1} = \frac{n-2i+1}{n+1} E(U_{i:n} - p_i)^c - \frac{i(n-i+1)}{(n+1)^2} E(U_{i:n}-p_i)^{c-1},$$

$$V_{c+1}(a,b;v) + v\, V_c(a,b;v) = W_c(a,b;v)$$

and

$$(a+b+c)\, W_{c+1}(a,b;v) = a - v(a+b) + c(1-2v)\, W_c(a,b;v) + cv(1-v)\, W_{c-1}(a,b;v),$$

where we take

$$c\, W_{c-1}(a,b;v) = -\frac{1}{B(a,b)} v^{a-1} (1-v)^{b-1} \qquad \text{if } c = 0$$
$$= I_v(a,b) \qquad \text{if } c = 1,$$

and

$$V_0(a,b;v) = \frac{a+b-1}{a-1} I_v(a-1,b);$$

$I_v(a,b)$ is Karl Pearson's incomplete beta function defined by

$$I_v(a,b) = \int_0^v \frac{1}{B(a,b)} u^{a-1}(1-u)^{b-1}\, du.$$

From (4.45) and (4.46), therefore, we will be able to get the bounds for the error involved in approximating $\mu_{i:n}$ by David and Johnson series up to 2m terms. Saw (1960) computed the bounds for the error involved in David and Johnson series as well as Plackett series. Upon comparing them, he observed that term for term the Plackett series converges a little more rapidly than does the David and Johnson series. However, as rightly pointed out by Saw (1960), the slight superiority of the Plackett series in respect of convergence is quite overshadowed by the computational advantages of the David and Johnson technique.

4.6. Sugiura's orthogonal inverse expansion

Sugiura (1962) applied an orthogonal inverse expansion approach to obtain approximations as well as bounds for the moments $\mu_{i:n}^{(k)}$ $(1 \le i \le n)$ for an arbitrary distribution. As explained in Section 3.1, by denoting $\{\phi_j(u)\}_{j=0}^{\infty}$ for a sequence of complete orthonormal functions in $L^2(0,1)$, we have for an arbitrary standardized distribution with $\mu = 0$ and $\sigma = 1$ (see Theorem 3.1)

$$\left|\mu_{i:n} - \sum_{j=1}^{k} a_j b_j\right| \le \left\{1 - \sum_{j=1}^{k} a_j^2\right\}^{1/2} \left\{\frac{B(2i-1,2n-2i+1)}{[B(i,n-i+1)]^2} - 1 - \sum_{j=1}^{k} b_j^2\right\}^{1/2}, \tag{4.47}$$

where

$$a_j = \int_0^1 F^{-1}(u)\, \phi_j(u)\, du$$

and

$$b_j = \int_0^1 \frac{1}{B(i,n-i+1)} u^{i-1}(1-u)^{n-i}\, \phi_j(u)\, du.$$

If the parent distribution is symmetic about zero, we may obtain improved bounds for $\mu_{i:n}$ as (see equation (3.27))

$$\left|\mu_{i:n} - \sum_{j=0}^{k} a_{2j+1} b_{2j+1}\right|$$

$$\le \left\{1 - \sum_{j=0}^{k} a_{2j+1}^2\right\}^{1/2} \left\{\frac{B(2i-1,2n-2i+1) - B(n,n)}{2[B(i,n-i+1)]^2} - \sum_{j=0}^{k} b_{2j+1}^2\right\}^{1/2}. \tag{4.48}$$

Further, if the parent distribution has a finite fourth moment, we may proceed similarly and obtain bounds for $\mu_{i:n}^{(2)}$ as

$$\left|\mu_{i:n}^{(2)} - \sum_{j=0}^{k} a_{2j}^{*} b_{2j}\right|$$

$$\leq \left\{E(X^4) - \sum_{j=0}^{k} a_{2j}^{*2}\right\}^{1/2} \left\{\frac{B(2i-1,2n-2i+1) + B(n,n)}{2[B(i,n-i+1)]^2} - \sum_{j=0}^{k} b_{2j}^2\right\}^{1/2}, \tag{4.49}$$

where

$$a_j^* = \int_0^1 \{F^{-1}(u)\}^2 \phi_j(u)\, du.$$

If we take the Legendre polynomials on (0,1) defined by

$$\phi_j(u) = \frac{\sqrt{2j+1}}{j!} \frac{d^j}{du^j} \{u^j(u-1)^j\}, \quad j = 0,1,2,...$$

$$= \sqrt{2j+1} \sum_{\ell=0}^{j} (-1)^{j-\ell} \binom{j}{\ell} \binom{j+\ell}{\ell} u^{\ell}$$

for the complete orthonormal system, then we have

$$a_j = \sqrt{2j+1} \sum_{\ell=0}^{j} (-1)^{j-\ell} \binom{j}{\ell} \binom{j+\ell}{\ell} \int_0^1 F^{-1}(u)\, u^{\ell}\, du$$

$$= \sqrt{2j+1} \sum_{\ell=0}^{j} (-1)^{j-\ell} \binom{j}{\ell} \binom{j+\ell}{\ell} \mu_{\ell+1:\ell+1}/(\ell+1), \tag{4.50}$$

$$a_j^* = \sqrt{2j+1} \sum_{\ell=0}^{j} (-1)^{j-\ell} \binom{j}{\ell} \binom{j+\ell}{\ell} \int_0^1 \{F^{-1}(u)\}^2 u^{\ell}\, du$$

$$= \sqrt{2j+1} \sum_{\ell=0}^{j} (-1)^{j-\ell} \binom{j}{\ell} \binom{j+\ell}{\ell} \mu_{\ell+1:\ell+1}^{(2)}/(\ell+1), \tag{4.51}$$

and

$$b_j = \sqrt{2j+1} \sum_{\ell=0}^{j} (-1)^{j-\ell} \binom{j}{\ell} \binom{j+\ell}{\ell} \frac{1}{B(i,n-i+1)} \int_0^1 u^{\ell+i-1}(1-u)^{n-i}\, du$$

$$= \sqrt{2j+1} \sum_{\ell=0}^{j} (-1)^{j-\ell} \binom{j}{\ell} \binom{j+\ell}{\ell} B(\ell+i,n-i+1)/B(i,n-i+1)$$

$$= \sqrt{2j+1} \sum_{\ell=0}^{j} (-1)^{j-\ell} \binom{j}{\ell} \binom{j+\ell}{\ell} \frac{i(i+1)\ldots(i+\ell-1)}{(n+1)(n+2)\ldots(n+\ell)}. \tag{4.52}$$

Realize that the Fourier coefficients a_j and a_j^* involved in deriving the bounds and approximations in (4.48) and (4.49) may be computed from equations (4.50) and (4.51) by using the first and the second moments of extreme order statistics in small sample sizes.

For the standard normal distribution, for example, we have

$$\int_0^1 F^{-1}(u)\, u\, du = \mu_{2:2}/2 = \frac{1}{2\sqrt{\pi}},$$

$$\int_0^1 F^{-1}(u)\, u^2\, du = \mu_{3:3}/2 = \frac{1}{2\sqrt{\pi}},$$

$$\int_0^1 F^{-1}(u)\, u^3\, du = \mu_{4:4}/4 = \frac{1}{4\pi\sqrt{\pi}} \operatorname{Cos}^{-1}(-1/3),$$

$$\int_0^1 F^{-1}(u)\, u^4\, du = \mu_{5:5}/5 = \frac{1}{2\sqrt{\pi}} \left\{1 - \frac{1}{\pi} \operatorname{Cos}^{-1}(-1/3)\right\}$$

and

$$\int_0^1 F^{-1}(u)\, u^5\, du = \mu_{6:6}/6 = 1.267206361/6;$$

the value of $\mu_{6:6}$ was taken from Teichroew (1956). Making use of the above results in (4.50), Sugiura (1962) computed the bounds and approximations for $\mu_{i:n}$ from (4.48) for n = 10 and 20. The results for the cases n = 10 and n = 20 are presented in Tables 4.1 and 4.2, respectively. From these tables, we see that the approximations are uniformly good for all i; this is not surprising since the orthogonal expansion gives the best approximation in the mean.

Table 4.1. Approximations and bounds for $\mu_{i:10}$ in the standard normal distribution*

i	k = 0	k = 1	k = 2	Exact value
10	1.38 ± 0.17	1.527 ± 0.015	1.5384 ± 0.0005	1.53875
9	1.08 ± 0.09	1.030 ± 0.035	1.0032 ± 0.0026	1.00136
8	0.77 ± 0.14	0.651 ± 0.008	0.6527 ± 0.0048	0.65606
7	0.46 ± 0.13	0.357 ± 0.028	0.3775 ± 0.0024	0.37576
6	0.15 ± 0.06	0.113 ± 0.016	0.1246 ± 0.0032	0.12267

Table 4.2. Approximations and bounds for $\mu_{i:20}$ in the standard normal distribution*

i	k = 0	k = 1	k = 2	Exact value
20	1.53 ± 0.35	1.796 ± 0.082	1.856 ± 0.016	1.867
19	1.37 ± 0.17	1.468 ± 0.067	1.433 ± 0.032	1.408
18	1.21 ± 0.15	1.186 ± 0.075	1.131 ± 0.012	1.131
17	1.05 ± 0.16	0.945 ± 0.057	0.907 ± 0.018	0.921
16	0.89 ± 0.18	0.739 ± 0.034	0.734 ± 0.022	0.745
15	0.73 ± 0.20	0.564 ± 0.037	0.587 ± 0.013	0.590
14	0.56 ± 0.21	0.413 ± 0.056	0.454 ± 0.008	0.448
13	0.40 ± 0.19	0.281 ± 0.065	0.325 ± 0.019	0.315
12	0.24 ± 0.14	0.163 ± 0.053	0.197 ± 0.020	0.187
11	0.081 ± 0.051	0.054 ± 0.021	0.066 ± 0.008	0.062

*Tables are reproduced with the permission of the author.

Sugiura (1964) has extended this approach to derive bounds and approximations for the product moment $\mu_{i,j:n}$; one may also refer to Mathai (1975, 1976) for some further generalizations.

4.7. Joshi's modified bounds and approximations

The orthogonal inverse expansion method of Sugiura (1962) discussed in the previous section requires that the population distribution has a finite variance. By a simple generalization of this method, Joshi (1969) has shown that, even with less stringent conditions, similar bounds and approximations may be derived for all finite moments of order statistics.

Theorem 4.1: Let $\phi_0 \equiv 1, \phi_1, \phi_2, \ldots$ be an orthonormal system in $L^2(0,1)$ and let $\mu^{(2)}_{2p+1:2p+2q+1}$ be finite for some integral $p,q \geq 0$. Then for $1 \leq i \leq n$

$$\left|\frac{B(p+i,q+n-i+1)}{B(i,n-i+1)}\mu_{p+i:p+q+n} - \sum_{j=0}^{k} a_j' b_j\right|$$

$$\leq \left\{B(2p+1,2q+1)\,\mu^{(2)}_{2p+1:2p+2q+1} - \sum_{j=0}^{k} a_j'^2\right\}^{1/2} \left\{\frac{B(2i-1,2n-2i+1)}{[B(i,n-i+1)]^2} - \sum_{j=0}^{k} b_j^2\right\}^{1/2}, \tag{4.53}$$

where

$$a_j' = \int_0^1 F^{-1}(u)\, u^p(1-u)^q\, \phi_j(u)\, du$$

and b_j is as defined in (4.52).

Proof. The result follows immediately upon applying (3.21) with

$$f(u) = F^{-1}(u)\, u^p(1-u)^q \text{ and } g(u) = \frac{1}{B(i,n-i+1)}\, u^{i-1}(1-u)^{n-i}.$$

Remark: Theorem 4.1 shows that the bounds and approximations for $\mu_{p+i:p+q+n}$ $(1 \leq i \leq n)$ may be obtained provided the moment $\mu^{(2)}_{2p+1:2p+2q+1}$ is finite. In terms of the pdf $f(x)$, a sufficient condition for this is (Sen, 1959)

$$\int_{-\infty}^{\infty} |x|^{2/(2p+1)} f(x)\, dx < \infty$$

for $q \geq p \geq 0$.

If the parent distribution is symmetric about zero, we may obtain some improved bounds and approximations as given in the following theorem.

Theorem 4.2: Let the parent distribution be symmetric about zero. Let $\phi_0 \equiv 1, \phi_1, \phi_2, \ldots$ be a complete orthonormal system in $L^2(0,1)$ such that $\phi_j(u) = (-1)^j\, \phi_j(1-u)$, $j = 1,2,\ldots$. If $\mu^{(2)}_{2m+1:4m+1}$ is finite for some integral $m \geq 0$, then for $1 \leq i \leq n$

$$\left|\frac{B(m+i,m+n-i+1)}{B(i,n-i+1)}\mu_{m+i:2m+n} - \sum_{j=0}^{k} a_{2j+1}' b_{2j+1}\right|$$

$$\le \left\{B(2m+1,2m+1)\,\mu^{(2)}_{2m+1:4m+1} - \sum_{j=0}^{k} a'^{2}_{2j+1}\right\}^{1/2}$$

$$\left\{\frac{B(2i-1,2n-2i+1) - B(n,n)}{2[B(i,n-i+1)]^2} - \sum_{j=0}^{k} b^2_{2j+1}\right\}^{1/2}, \tag{4.54}$$

where a'_j is as in Theorem 4.1, with $p = q = m$.

Moreover, if $\mu^{(4)}_{2m+1:4m+1}$ is finite for some integral $m \ge 0$, then for $1 \le i \le n$

$$\left|\frac{B(m+i,m+n-i+1)}{B(i,n-i+1)}\,\mu^{(2)}_{m+i:2m+n} - \sum_{j=0}^{k} a''_{2j}\, b_{2j}\right|$$

$$\le \left\{B(2m+1,2m+1)\,\mu^{(4)}_{2m+1:4m+1} - \sum_{j=0}^{k} a''^{2}_{2j}\right\}^{1/2}$$

$$\left\{\frac{B(2i-1,2n-2i+1) + B(n,n)}{2[B(i,n-i+1)]^2} - \sum_{j=0}^{k} b^2_{2j}\right\}^{1/2}, \tag{4.55}$$

where

$$a''_j = \int_0^1 \{F^{-1}(u)\}^2\, u^m(1-u)^m\, \phi_j(u)\, du.$$

Proof. To prove (4.54), we take

$$f(u) = F^{-1}(u)\, u^m(1-u)^m \text{ and } g(u) = \frac{1}{B(i,n-i+1)}\, u^{i-1}(1-u)^{n-i},$$

apply (3.23) with $j = \{1,3,\ldots,2k+1,0,2,4,6,\ldots\}$ and use the results that

$$a'_{2j} = \int_0^1 F^{-1}(u)\, u^m(1-u)^m\, \phi_{2j}(u)\, du = 0$$

and

$$\sum_{j=0}^{\infty} b^2_{2j} = \int_0^1 \frac{1}{4}\left\{g(u) + g(1-u)\right\}^2 du$$

$$= \frac{B(2i-1,2n-2i+1) + B(n,n)}{2[B(i,n-i+1)]^2}.$$

Similarly, to prove (4.55), we take

$$f(u) = \{F^{-1}(u)\}^2\, u^m(1-u)^m \text{ and } g(u) = \frac{1}{B(i,n-i+1)}\, u^{i-1}(1-u)^{n-i},$$

apply (3.23) with $J = \{0,2,4,\ldots,2k,1,3,5,7,\ldots\}$ and use the results that

$$a''_{2j+1} = \int_0^1 \{F^{-1}(u)\}^2\, u^m(1-u)^m\, \phi_{2j+1}(u)\, du = 0$$

and

$$\sum_{j=0}^{\infty} b^2_{2j+1} = \int_0^1 \frac{1}{4}\{g(u) - g(1-u)\}^2\, du$$

$$= \frac{B(2i-1,2n-2i+1) - B(n,n)}{2[B(i,n-i+1)]^2}.$$

If we now take the sequence of Legendre polynomials on (0,1) for the complete orthonormal system

$\{\phi_j(u)\}_{j=0}^{\infty}$, then we have

$$a_j' = \sqrt{2j+1} \sum_{\ell=0}^{j} (-1)^{j-\ell} \binom{j}{\ell} \binom{j+\ell}{\ell} \int_0^1 F^{-1}(u)\, u^{\ell+m}(1-u)^m \, du$$

$$= \sqrt{2j+1} \sum_{\ell=0}^{j} (-1)^{j-\ell} \binom{j}{\ell} \binom{j+\ell}{\ell} B(\ell+m+1,m+1)\, \mu_{\ell+m+1:\ell+2m+1}, \tag{4.56}$$

$$a_j'' = \sqrt{2j+1} \sum_{\ell=0}^{j} (-1)^{j-\ell} \binom{j}{\ell} \binom{j+\ell}{\ell} \int_0^1 \{F^{-1}(u)\}^2\, u^{\ell+m}(1-u)^m \, du$$

$$= \sqrt{2j+1} \sum_{\ell=0}^{j} (-1)^{j-\ell} \binom{j}{\ell} \binom{j+\ell}{\ell} B(\ell+m+1,m+1)\, \mu^{(2)}_{\ell+m+1:\ell+2m+1}, \tag{4.57}$$

and b_j is as in equation (4.52).

For the Cauchy distribution with pdf

$$f(x) = \frac{1}{\pi(1+x^2)}, \qquad -\infty < x < \infty,$$

Barnett (1966) has shown that $\mu_{i:n}^{(i')}$ is finite for all $i' < i \le n-i'$. In particular, $\mu_{3:5}^{(2)}$ is finite and hence (4.54) is applicable with m = 1. We may thus get bounds for $\mu_{i+1:n+2}$, $1 \le i \le n$, which are all the finite moments in this case. For the case m = 1, we note from (4.56) that a_j' is a linear function of the moments $\mu_{\ell+1:\ell+2}$ in small samples which are usually available; see, for example, Tietjen et al. (1977) for the normal distribution, and Barnett (1966) and Rider (1960) for the Cauchy distribution. Making use of Theorem 4.2, Joshi (1969) computed the bounds and approximations for $\mu_{i:n}$ $(2 \le i \le n-1)$ when n = 10 and 20 for the standard normal and Cauchy distributions. These are presented in Tables 4.3 – 4.5. From these tables, we see that the approximations and bounds are remarkably good for all i $(2 \le i \le n-1)$, especially for the Cauchy distribution.

As mentioned earlier, we can obtain several different sequences of bounds all converging to $\mu_{i+m:n+2m}$ by choosing different values of m and n in (4.54). Unfortunately, there is no theoretical way of determining the choice of m, for which the difference between the true value of $\mu_{i+m:n+2m}$ and its approximate value is minimum for a fixed value of k. After performing some numerical computations for the normal distribution with n + 2m = 20 and 50, Joshi (1969) observed that the case when m = 1 gives better results than the cases when m = 0 or m = 2 for the second approximation (that is, when k = 1).

Table 4.3. Approximations and bounds for $\mu_{i:10}$ in the standard normal distribution*

i	k = 0	k = 1	Exact value
9	1.2994 ± 0.2982	1.0041 ± 0.0028	1.00136
8	0.5304 ± 0.1325	0.6509 ± 0.0053	0.65606
7	0.2475 ± 0 1355	0.3788 ± 0.0031	0.37576
6	0.0743 ± 0.0567	0.1249 ± 0.0025	0.12267

Table 4.4. Approximations and bounds for $\mu_{i:10}$ in the Cauchy distribution*

i	k = 0	k = 1	Exact value
9	3.0529 ± 0.0727	2.9822 ± 0.0015	2.9814
8	1.2461 ± 0.0323	1.2749 ± 0.0028	1.2755
7	0.5815 ± 0.0330	0.6129 ± 0.0016	0.6132
6	0.1745 ± 0.0138	0.1866 ± 0.0013	0.1866

Table 4.5. Approximations and bounds for $\mu_{i:20}$ in the standard normal and Cauchy distributions*

i	Normal distribution		Cauchy distribution	
	k = 1	Exact value	k = 1	Exact value
19	1.4693 ± 0.0619	1.40760	6.2705 ± 0.0324	6.2648
18	1.1023 ± 0.0310	1.13095	3.0287 ± 0.0162	3.0293
17	0.9002 ± 0.0225	0.92098	1.9128 ± 0.0118	1.9140
16	0.7390 ± 0.0116	0.74538	1.3259 ± 0.0060	1.3268
15	0.5940 ± 0.0063	0.59030	0.9480 ± 0.0033	0.9484
14	0.4570 ± 0.0094	0.44833	0.6718 ± 0.0049	0.6720
13	0.3241 ± 0.0115	0.31493	0.4506 ± 0.0060	0.4506
12	0.1937 ± 0.0095	0.18696	0.2600 ± 0.0050	0.2599
11	0.0644 ± 0.0037	0.06200	0.0851 ± 0.0019	0.0850

*Tables are reproduced with the permission of the author

4.8. Joshi and Balakrishnan's improved bounds for extremes

By noting that the bounds for extreme order statistics obtained by Sugiura's method may not be sharp and that the moments are usually tabulated for some typical values of n in case of large sample sizes, Joshi and Balakrishnan (1983) and Balakrishnan and Joshi (1985) have developed a method of obtaining improved bounds and approximations for moments of extreme order statistics by making use of few neighbouring tabulated values. These bounds are of similar type as those of Sugiura (1962) discussed in Section 6.

To fix the ideas, let $\mu_{n:n}$ be the moment for which bounds are required and m and p be any two integers in the neighbourhood of n for which $\mu_{m:m}$ and $\mu_{p:p}$ are tabulated. We can then write

$$\mu_{n:n} - c\,\mu_{m:m} - d\,\mu_{p:p} = \int_0^1 F^{-1}(u)\,\{nu^{n-1} - c\,mu^{m-1} - d\,pu^{p-1}\}\,du,$$

where c and d are constants so chosen that the error bound by applying the inequality in (3.21) for a (k+l) – term approximation is as small as possible.

Then, upon applying (3.21) with

$$f(u) = F^{-1}(u) \text{ and } g(u) = nu^{n-1} - c\,mu^{m-1} - d\,pu^{p-1}$$

and the sequence of Legendre polynomials as the orthonormal system $\{\phi_k(u)\}_{k=0}^{\infty}$, we obtain

$$\left|\mu_{n:n} - c\,\mu_{m:m} - d\,\mu_{p:p} - \sum_{j=0}^{k} a_j b_j\right|$$

$$\le \left\{E(X^2) - \sum_{j=0}^{k} a_j^2\right\}^{1/2} \left[\frac{n^2}{2n-1} + \frac{c^2m^2}{2m-1} + \frac{d^2p^2}{2p-1} - \frac{2cmn}{m+n-1} - \frac{2dpn}{p+n-1} + \frac{2cdmp}{m+p-1} - \sum_{j=0}^{k} b_j^2\right]^{1/2}, \tag{4.58}$$

where

$$a_j = \int_0^1 F^{-1}(u)\,\phi_j(u)\,du \tag{4.59}$$

and

$$b_j = \int_0^1 \left\{nu^{n-1} - c\,mu^{m-1} - d\,pu^{p-1}\right\} \phi_j(u)\,du$$

$$= b_{j,n} - c\,b_{j,m} - d\,b_{j,p}; \tag{4.60}$$

we have

$$\phi_j(u) = \frac{\sqrt{2j+1}}{j!}\frac{d^j}{du^j}\{u^j(u-1)^j\}$$

$$= \sqrt{2j+1}\sum_{\ell=0}^{j} (-1)^{j-\ell}\binom{j}{\ell}^2 u^{\ell}(1-u)^{j-\ell} \tag{4.61}$$

by applying Leibniz rule and, hence,

$$b_{j,s} = \int_0^1 su^{s-1}\,\phi_j(u)\,du$$

$$= \sqrt{2j+1}\sum_{\ell=0}^{j} (-1)^{j-\ell}\binom{j}{\ell}^2 \int_0^1 su^{s+\ell-1}(1-u)^{j-\ell}\,du$$

$$= \sqrt{2j+1}\sum_{\ell=0}^{j} (-1)^{j-\ell}\binom{j}{\ell}\binom{s+\ell-1}{\ell}\Big/\binom{s+j}{s}$$

$$= \sqrt{2j+1}\,\frac{(s-1)(s-2)\dots(s-j)}{(s+1)(s+2)\dots(s+j)}. \tag{4.62}$$

Equation (4.62) follows immediately by considering the identity $(1-t)^j(1-t)^{-s} = (1-t)^{j-s}$ and equating the coefficients of t^j on both sides. Noting now that only the second factor on the RHS of the inequality in (4.58) contains c and d and substituting for b_j from equation (4.60) and minimizing the resulting expression with respect to c and d, we obtain the optimal values of c and d for (k+l) – term approximation as

$$c_{opt} = (R_{n,p}R_{m,p} - R_{n,m}R_{p,p})/(R_{m,p}^2 - R_{m,m}R_{p,p})$$

and

$$d_{opt} = (R_{n,m}R_{m,p} - R_{n,p}R_{m,m})/(R^2_{m,p} - R_{m,m}R_{p,p}),$$

where

$$R_{a,b} = \sum_{j=0}^{k} b_{j,a}b_{j,b} - \frac{ab}{a+b-1}.$$

Making use of these optimal values for c and d in (4.58), we may get an approximate value for $\mu_{n:n}$ as $c_{opt}\,\mu_{m:m} + d_{opt}\,\mu_{p:p} + \sum_{j=0}^{k} a_j b_j$ along with a bound given by the RHS of (4.58).

If the parent distribution is symmetric about zero, we may obtain some improved bounds and approximations as given in the following theorem.

Theorem 4.3: For positive integers a and b, let

$$H(a,b) = \frac{ab}{2}\left\{\frac{1}{a+b-1} - B(a,b)\right\},$$

where B(a,b) is the complete beta function. Then for symmetric distributions with mean 0 and variance 1, we have for $k \geq 0$

$$\left|\mu_{n:n} - c\,\mu_{m:m} - d\,\mu_{p:p} - \sum_{j=0}^{k} a_{2j+1}b_{2j+1}\right|$$

$$\leq \left\{1 - \sum_{j=0}^{k} a^2_{2j+1}\right\}^{1/2} \Bigg\{H(n,n) + c^2\,H(m,m) + d^2\,H(p,p) - 2c\,H(m,n)$$

$$- 2d\,H(p,n) + 2cd\,H(m,p) - \sum_{j=0}^{k} b^2_{2j+1}\Bigg\}^{1/2}, \tag{4.63}$$

where a_j and b_j are as given in (4.59) and (4.60), respectively, and the optimal values of c and d are given by

$$c_{opt} = (S_{n,p}S_{m,p} - S_{n,m}S_{p,p})/(S^2_{m,p} - S_{m,m}S_{p,p})$$

and

$$d_{opt} = (S_{n,m}S_{m,p} - S_{n,p}S_{m,m})/(S^2_{m,p} - S_{m,m}S_{p,p}),$$

with

$$S_{a,b} = \sum_{j=0}^{k} b_{2j+1,a}\,b_{2j+1,b} - H(a,b).$$

Proof. To prove (4.63), we take

$$f(u) = F^{-1}(u) \quad \text{and} \quad g(u) = nu^{n-1} - c\,mu^{m-1} - d\,pu^{p-1},$$

apply (3.23) with J = {1,3,...,2k+1,0,2,4,6,...}, and use the results that

$$a_{2j} = \int_0^1 F^{-1}(u)\,\phi_{2j}(u)\,du = 0$$

and

$$\sum_{j=0}^{\infty} b^2_{2j} = \int_0^1 \left\{\frac{g(u) + g(1-u)}{2}\right\}^2 du$$

$$= G(n,n) + c^2\, G(m,m) + d^2\, G(p,p) - 2c\, G(m,n) - 2d\, G(p,n) + 2cd\, G(m,p),$$

where G(a,b), for positive integers a and b, is

$$G(a,b) = \frac{ab}{2}\left\{\frac{1}{a+b-1} + B(a,b)\right\}. \tag{4.64}$$

Proceeding similarly, we can prove the following theorem which yields bounds and approximations for the second moment of $X_{n:n}$.

Theorem 4.4: For positive integers a and b, let G(a,b) be as defined in (4.64). Then for distributions symmetric about zero with variance 1 and finite $E(X^4)$, we have for $k \ge 0$

$$\left|\mu_{n:n}^{(2)} - c\,\mu_{m:m}^{(2)} - d\,\mu_{p:p}^{(2)} - \sum_{j=0}^{k} a_{2j}^{*} b_{2j}\right|$$

$$\le \left\{E(X^4) - \sum_{j=0}^{k} a_{2j}^{*2}\right\}^{1/2} \left\{G(n,n) + c^2\, G(m,m) + d^2\, G(p,p) - 2c\, G(m,n)\right.$$

$$\left. - 2d\, G(p,n) + 2cd\, G(m,p) - \sum_{j=0}^{k} b_{2j}^2\right\}^{1/2}, \tag{4.65}$$

where, as before,

$$a_j^* = \int_0^1 \{F^{-1}(u)\}^2\, \phi_j(u)\, du$$

and the optimal values of c and d are given by

$$c_{opt} = (T_{n,p}T_{m,p} - T_{n,m}T_{p,p})/(T_{m,p}^2 - T_{m,m}T_{p,p})$$

and

$$d_{opt} = (T_{n,m}T_{m,p} - T_{n,p}T_{m,m})/(T_{m,p}^2 - T_{m,m}T_{p,p}),$$

with

$$T_{a,b} = \sum_{j=0}^{k} b_{2j,a}\, b_{2j,b} - G(a,b).$$

By making use of the expression of the Legendre polynomials given in equation (4.61), Joshi and Balakrishnan (1983) obtained an expression for the Fourier coefficients a_j as

$$a_j = \frac{\sqrt{2j+1}}{j+1} \sum_{\ell=0}^{j} (-1)^{j-\ell} \binom{j}{\ell} \mu_{\ell+1:j+1}, \quad j \ge 0. \tag{4.66}$$

Equation (4.66) expresses a_j as a linear combination of $\mu_{1:j+1}, \mu_{2:j+1}, \ldots, \mu_{j+1:j+1}$ with coefficients which are not too large unlike the expression given in (4.50). Similarly, we may write the Fourier coefficients a_j^* as

$$a_j^* = \frac{\sqrt{2j+1}}{j+1} \sum_{\ell=0}^{j} (-1)^{j-\ell} \binom{j}{\ell} \mu_{\ell+1:j+1}^{(2)}, \quad j \ge 0. \tag{4.67}$$

By making use of the means of standard normal order statistics tabulated by Yamauti (1972) and the table of second moments prepared by Teichroew (1956), Joshi and Balakrishnan (1983) have computed the coefficients a_j and a_j^* for the standard normal case; these values are presented in Tables 4.6 and 4.7, respectively.

For large values of n, m and p, the beta term in the function H(a,b) in (4.63) becomes negligible and hence can be ignored. This reduces (4.63) to a much simpler form. Then, Joshi and Balakrishnan (1983) applied this simplified form of Theorem 4.3 to compute approximation and bound for $\mu_{n:n}$ for n = 100 (100) 1000 by taking m = 400, p = 1000 and k = 13, and these are presented in Table 4.8. For

Table 4.6. Fourier coefficients a_j for the standard normal distribution

j	a_j
1	0.9772050237
3	0.1830082402
5	0.0816989763
7	0.0477293675
9	0.0318804432
11	0.0230790883
13	0.0176307593
15	0.0139963491
17	0.0114376017
19	0.0095600272
21	0.0081467944
23	0.0070483945
25	0.0064146017
27	0.0059105768

Table 4.7. Fourier coefficients a_j^* for the standard normal distribution

j	a_j^*
0	1.0000000000
2	1.2328088881
4	0.5211245856
6	0.3045144707
8	0.2055889844
10	0.1507706975
12	0.1166778189
14	0.0937736622
16	0.0775095317
18	0.0654684404

comparison purposes, we also present in this table the approximation and bound obtained by Sugiura's method with k = 13 and the tabled values taken from the tables of Harter (1961) and Tippett (1925). From this table, we see that the bounds and approximations obtained from Theorem 4.3 are considerably better than those of Sugiura.

For large values of n, m and p, the beta term in the function G(a,b) in (4.65) becomes negligible and hence can be ignored. By using this simplified form of Theorem 4.4, Balakrishnan (1980) computed the approximation and bound for $\mu_{n:n}^{(2)}$ for n = 75 (5) 120 by taking m = 80, p = 120 and k = 9, and these are presented in Table 4.9. Once again, for the purpose of comparison we present in this table the approximation and bound computed by Sugiura's method with k = 9 and the tabled values taken from Borenius (1966). From this table, we see once again that the bounds and approximations obtained from Theorem 4.4 to be better than those of Sugiura (1962).

Balakrishnan and Joshi (1985) have applied successfully Joshi's method to extend the results presented in this section in order to derive improved bounds and approximations for the moments of the second largest order statistic. It should be remarked here that the Fourier coefficients a_j given in Table 4.6 can be used to approximate the sum of squares of normal scores, viz., $S = \sum_{i=1}^{n} \mu_{i:n}^2$; see, for example, Ruben (1954), Saw and Chow (1966), and Balakrishnan (1984).

Table 4.8. Approximations and bounds for $\mu_{n:n}$ in the standard normal distribution

n	Sugiura's bound	Bound from (4.63)	Exact value
100	2.507593 ± 0.000016	2.507599 ± 0.000007	2.50759
200	2.745307 ± 0.001456	2.746076 ± 0.000315	2.74604
300	2.873277 ± 0.007087	2.877765 ± 0.000536	2.87777
400	2.956663 ± 0.016230	2.968180 ± 0.000000	2.96818
500	3.015873 ± 0.027329	3.036800 ± 0.001002	3.03670
600	3.060209 ± 0.039313	3.091950 ± 0.001950	3.09170
700	3.094683 ± 0.051573	3.137915 ± 0.002465	3.13755
800	3.122263 ± 0.063778	3.177185 ± 0.002348	3.17679
900	3.144831 ± 0.075755	3.211338 ± 0.001526	3.21105
1000	3.163642 ± 0.087418	3.241440 ± 0.000000	3.24144

Table 4.9. Approximations and bounds for $\mu_{n:n}^{(2)}$ in the standard normal distribution

n	Sugiura's bound	Bound from (4.65)	Exact value
75	5.9689676 ± 0.0031344	5.9706607 ± 0.0000071	5.9706671
80	6.0802196 ± 0.0045483	6.0827371 ± 0.0000000	6.0827371
85	6.1847147 ± 0.0063162	6.1882830 ± 0.0000568	6.1882793
90	6.2831558 ± 0.0084588	6.2880338 ± 0.0000832	6.2880179
95	6.3761356 ± 0.0109903	6.3825878 ± 0.0001029	6.3825623
100	6.4641592 ± 0.0139167	6.4724659 ± 0.0000935	6.4724306
105	6.5476619 ± 0.0172393	6.5581089 ± 0.0001423	6.5580668
110	6.6270211 ± 0.0209553	6.6398926 ± 0.0001337	6.6398545
115	6.7025678 ± 0.0250577	6.7181548 ± 0.0000991	6.7181269
120	6.7745939 ± 0.0295364	6.7931759 ± 0.0000000	6.7931759

The coefficients J_j defined by Saw and Chow in their equation (14) for the purpose of approximating S are related to the Fourier coefficients a_j; in fact, it can be shown that $a_j^2 = (2j+1)\, J_j^2$. It is also of interest to mention here that Royston (1982) has developed an efficient algorithm for computing the exact and approximate values of expected normal order statistics.

Exercises

1. (Ruben (1954); Saw and Chow (1966); Balakrishnan (1984)). With a_j as given in Table 4.6, show for the standard normal distribution that

$$\frac{1}{n} S = \frac{1}{n} \sum_{i=1}^{n} \mu_{i:n}^2 = \sum_{j=0}^{2k-1} a_j^2 \frac{n!\ (n-1)!}{(n+j)!(n-j-1)!} + R_{2k-1},$$

where $a_{2j} = 0$ ($j = 0,1,2,\ldots$) and R_{2k-1} is the error term. If the upper and lower bounds for R_{2k-1} are denoted by R^*_{2k-1} and R_{*2k-1}, respectively, then show that

$$R^*_{2k-1} - R_{*2k-1} = \frac{(n-1)(n-2)\ldots(n-2k-1)}{(n+1)(n+2)\ldots(n+2k+1)} \left\{1 - \sum_{j=0}^{2k+1} a_j^2\right\}$$

which can be evaluated for any given value of n and k.

2. (Balakrishan and Joshi (1985)). (i) For positive integers a and b, let the function H(a,b) be as defined in Theorem 4.3. Then for standardized symmetric distribution with finite $\mu^{(2)}_{3:5}$, show that for $k \geq 0$

$$\left|\frac{n}{(n+1)(n+2)} \mu_{n+1:n+2} - c \frac{m}{(m+1)(m+2)} \mu_{m+1:m+2} - d \frac{p}{(p+1)(P+2)} \mu_{p+1:p+2} - \sum_{j=0}^{k} a'_{2j+1} b_{2j+1}\right|$$

$$\leq \left\{\frac{1}{30} \mu^{(2)}_{3:5} - \sum_{j=0}^{k} a'^2_{2j+1}\right\}^{1/2} \left\{H(n,n) + c^2 H(m,m) + d^2 H(p,p) - 2c\, H(m,n) - 2d\, H(p,n) + 2cd\, H(m,p) - \sum_{j=0}^{k} b^2_{2j+1}\right\}^{1/2},$$

where

$$a'_j = \int_0^1 F^{-1}(u)\, u(1-u)\, \phi_j(u)\, du$$

and b_j is as defined in (4.60), and the optimal values of c and d are as given in Theorem 4.3.

(ii) With Fourier coefficients a_j as defined in equation (4.59), show for distributions symmetric about zero that for $j = 1,3,5,\ldots$

$$-4a'_j = \frac{j(j-1)}{(2j-3)^{1/2}(2j-1)(2j+1)^{1/2}} a_{j-2} + \frac{(j+1)(j+2)}{(2j+1)^{1/2}(2j+3)\ (2j+5)^{1/2}} a_{j+2} + \left\{\frac{(j+1)^2}{(2j+1)(2j+3)} + \frac{j^2}{(2j-1)(2j+1)} - 1\right\} a_j$$

with the convention that $a_{-1} = 0$.

3. (Balakrishnan and Joshi (1985)). For positive integers a and b, let

$$F(a,b) = ab(a-1)(b-1)\left\{\frac{1}{(a+b-1)(a+b-2)(a+b-3)} - \frac{1}{2}B(a,b)\right\},$$

where B(a,b) is the complete beta function. Then for symmetric distributions with mean 0 and variance 1, show that for $k \geq 0$

$$\left|\mu_{n-1:n} - c\,\mu_{m-1:m} - d\,\mu_{p-1:p} - \sum_{j=0}^{k} a_{2j+1} b'_{2j+1}\right|$$

$$\leq \left\{1 - \sum_{j=0}^{k} a_{2j+1}^2\right\}^{1/2} \left\{F(n,n) + c^2\,F(m,m) + d^2\,F(p,p) - 2c\,F(m,n)\right.$$

$$\left. - 2d\,F(p,n) + 2cd\,F(m,p) - \sum_{j=0}^{k} b'^2_{2j+1}\right\}^{1/2},$$

where a_j is as defined in equation (4.59),

$$b'_j = \int_0^1 \left\{(n-1)u^{n-2}(1-u) - c\,m(m-1)u^{m-2}(1-u) - d\,p(p-1)u^{p-2}(1-u)\right\} \phi_j(u)du$$

$$= \left\{n\,b_{j,n-1} - (n-1)\,b_{j,n}\right\} - c\left\{m\,b_{j,m-1} - (m-1)\,b_{j,m}\right\} - d\left\{p\,b_{j,p-1} - (p-1)\,b_{j,p}\right\}$$

with $b_{j,s}$ as in (4.62), and the optimal values of c and d are given by

$$c_{opt} = (\upsilon_{m,p}\upsilon_{n,p} - \upsilon_{m,n}\upsilon_{p,p})/(\upsilon^2_{m,p} - \upsilon_{m,m}\upsilon_{p,p}),$$

with

$$\upsilon_{a,b} = F(a,b) - \sum_{j=0}^{k} \left\{a\,b_{2j+1,a-1} - (a-1)\,b_{2j+1,a}\right\}\left\{b\,b_{2j+1,b-1} - (b-1)\,b_{2j+1,b}\right\}.$$

4. For the case of the standard normal distribution, by discarding the beta terms in the inequalities in Exercises 2 and 3, compute the bounds and approximation for $\mu_{n+1:n+2}$ for n = 48 (50) 398 by taking m = 248, p = 398 and k = 8. Compare these with the bounds and approximations of the same order obtained by Sugiura's and Joshi's methods along with the tabled values taken from the tables of Harter (1961) and comment.

5. Let $L^2(0,1)$ and $L^2(R)$ be the spaces of all square integrable functions in (0,1) and the square $R = \{(u,v):\ 0 \leq u,\ v \leq 1\}$, respectively. If $\{\phi_k(u)\}_{k=0}^{\infty}$ is a complete orthonormal system in $L^2(0,1)$, then prove that $\{\phi_k(u)\,\phi_\ell(v)\}_{k,\ell=0}^{\infty}$ is a complete orthonormal system in $L^2(R)$.

6. Decompose $L^2(R)$ into the following four subspaces:

$$L^2_{e,e}(R) = \left\{f(u,v)\,|\,f \in L^2(R) \text{ and } f(u,v) = f(1-u,v) = f(u,1-v)\right\},$$

$$L^2_{e,o}(R) = \left\{f(u,v)\,|\,f \in L^2(R) \text{ and } f(u,v) = f(1-u,v) = -f(u,1-v)\right\},$$

$$L^2_{o,e}(R) = \left\{f(u,v)\,|\,f \in L^2(R) \text{ and } f(u,v) = -f(1-u,v) = f(u,1-v)\right\},$$

$$L^2_{o,o}(R) = \left\{f(u,v)\,|\,f \in L^2(R) \text{ and } f(u,v) = -f(1-u,v) = -f(u,1-v)\right\}.$$

Let $\{\phi_k(u)\}_{k=0}^{\infty}$ be any complete orthonormal system in $L^2(0,1)$ satisfying $\phi_k(u) = (-1)^k \phi_k(1-u)$. Then, show that $\{\phi_{2k}(u)\,\phi_{2\ell}(v)\}_{k,\ell=0}^{\infty} \in L^2_{e,e}(R)$, $\{\phi_{2k}(u)\,\phi_{2\ell+1}(v)\}_{k,\ell=0}^{\infty} \in L^2_{e,o}(R)$, $\{\phi_{2k+1}(u)\,\phi_{2\ell}(v)\}_{k,\ell=0}^{\infty} \in L^2_{o,e}(R)$, $\{\phi_{2k+1}(u)\,\phi_{2\ell+1}(v)\}_{k,\ell=0}^{\infty} \in L^2_{o,o}(R)$ and that each subsystem is complete and orthonormal in the corresponding subspace.

7. Suppose that F(x) is the cdf of a random variable X and that $E(X^2)$ is finite. Then, show that $F^{-1}(u)F^{-1}(v) \in L^2(R)$. If, further, the distribution is symmetric about 0, then show that $F^{-1}(u)F^{-1}(v) \in L^2_{o,o}(R)$.

8. (Sugiura (1964)). (i) Let $\{\phi_k(u)\}_{k=0}^{\infty}$, with $\phi_0 \equiv 1$, be a sequence of complete orthonormal functions in $L^2(0,1)$. Then for an arbitrary standardized distribution with mean 0 and variance 1, show that for $1 \le i < j \le n$ and $t \ge 1$

$$\left|\mu_{i,j:n} - \frac{1}{2}\sum_{k,\ell=1}^{t} a_k a_\ell \left[b_{k,\ell} + b_{\ell,k}\right]\right|$$

$$\le \left\{1 - \sum_{k,\ell=1}^{t} a_k^2 a_\ell^2\right\}^{1/2} \left\{\frac{B(2i-1,2j-2i-1,2n-2j+1)}{2[B(i,j-i,n-j+1)]^2} - \frac{B(2i-1,2n-2i+1)}{2[B(i,n-i+1)]^2}\right.$$

$$\left. - \frac{B(i+j-1,2n-i-j+1)}{B(i,n-i+1)B(j,n-j+1)} - \frac{B(2j-1,2n-2j+1)}{2[B(j,n-j+1)]^2} + 1 - \frac{1}{4}\sum_{k,\ell=1}^{t}\left[b_{k,\ell} + b_{\ell,k}\right]^2\right\}^{1/2},$$

where

$$a_k = \int_0^1 F^{-1}(u)\,\phi_k(u)\,du,$$

$$b_{k,\ell} = \frac{1}{B(i,j-i,n-j+1)} \iint_{0<u<v<1} u^{i-1}(v-u)^{j-i-1}(1-v)^{n-j}\,\phi_k(u)\phi_\ell(v)\,du\,dv$$

and

$$B(a,b,c) = \frac{\Gamma(a)\,\Gamma(b)\,\Gamma(c)}{\Gamma(a+b+c)}, \quad (a,b,c > 0).$$

(ii) For an arbitrary standardized distribution with mean 0 and variance 1, show that for $1 \le i < j \le n$

$$\left|\mu_{i,j:n}\right| \le \left\{\frac{B(2i-1,2j-2i-1,2n-2j+1)}{2[B(i,j-i,n-j+1)]^2} - \frac{B(2i-1,2n-2j+1)}{2[B(i,n-i+1)]^2} - \frac{B(2j-1,2n-2j+1)}{2[B(j,n-j+1)]^2}\right.$$

$$\left. - \frac{B(i+j-1,2n-i-j+1)}{B(i,n-i+1)\,B(j,n-j+1)} + 1\right\}^{1/2} = \delta \text{ (say)},$$

and that equality holds if and only if

$$F^{-1}(u)\,F^{-1}(v) = \pm\frac{1}{\delta}\left\{1 + g(u,v) - \frac{u^{i-1}(1-u)^{n-i} + v^{i-1}(1-v)^{n-i}}{2\,B(i,n-i+1)}\right.$$

$$\left. - \frac{u^{j-1}(1-u)^{n-j} + v^{j-1}(1-v)^{n-j}}{2B(j,n-j+1)}\right\},$$

where

$$g(u,v) = \frac{1}{2B(i,j-i,n-j+1)}\,u^{i-1}(v-u)^{j-i-1}(1-v)^{n-j}, \quad 0 < u < v < 1$$

$$= \frac{1}{2B(i,j-i,n-j+1)}\, v^{i-1}(u-v)^{j-i-1}(1-u)^{n-j},\ 0 < v < u < 1.$$

9. By defining the function I(a,b,c,d,e) as

$$I(a,b,c,d,e) = \iint_{\substack{0<u<v<1\\ u+v<1}} u^{a-1}v^{b-1}(1-v)^{c-1}(v-u)^{d-1}(1-u-v)^{e-1}\,du\,dv$$

for positive integers a,b,c,d,e, show that

$$I(a,b,c,d,e) = \sum_{k=0}^{b-1}\sum_{\ell=0}^{c-1} \binom{b-1}{k}\binom{c-1}{\ell} B(d+k,e+\ell,a+b+c-k-\ell-2)\, 2^{-(a+b+c-k-\ell-2)}$$

and

$$I(a,b,c,d,e) = I(a,c,b,e,d).$$

10. (Sugiura (1964)). (i) Let $\{\phi_k(u)\}_{k=0}^{\infty}$, with $\phi_0 \equiv 1$, be a sequence of complete orthonormal functions in $L^2(0,1)$. Then for an arbitrary symmetric distribution with mean 0 and variance 1, show that for $1 \le i < j \le n$ and $t \ge 0$

$$\left|\mu_{i,j:n} - \frac{1}{2}\sum_{k,\ell=0}^{t} a_{2k+1}\, a_{2\ell+1}\left[b_{2k+1,2\ell+1} + b_{2\ell+1,2k+1}\right]\right|$$

$$\le \left\{1 - \sum_{k,\ell=0}^{t} a_{2k+1}^2\, a_{2\ell+1}^2\right\}^{1/2} \left\{K_1 - K_2 - \frac{1}{4}\sum_{k,\ell=0}^{t} \left[b_{2k+1,2\ell+1} + b_{2\ell+1,2k+1}\right]^2\right\}^{1/2},$$

where

$$K_1 = \frac{B(2i-1,2j-2i-1,2n-2j+1)+B(n+i-j,n+i-j,2j-2i-1)}{8[B(i,j-i,n-j+1)]^2},$$

$$K_2 = \frac{1}{8[B(i,j-i,n-j+1)]^2}\Big\{I(2i-1,n-j+1,n-j+1,j-i,j-i)$$

$$+ 2I(n+i-j,i,n-j+1,j-i,j-i) + I(2n-2j+1,i,i,j-i,j-i)\Big\},$$

a_k and $b_{k,\ell}$ as defined in Exercise 8, and I(a,b,c,d,e) as defined in Exercise 9.

(ii) For an arbitrary symmetric distribution with mean 0 and variance 1, show that for $1 \le i < j \le n$

$$|\mu_{i,j:n}| \le \frac{1}{2\sqrt{2}\,B(i,j-i,n-j+1)} \Big\{B(2i-1,2j-2i-1,2n-2j+1) + B(n+i-j,n+i-j,2j-2i-1) - I(2i-1,n-j+1,n-j+1,j-i,j-i) - 2I(n+i-j,i,n-j+1,j-i,j-i) - I(2n-2j+1,i,i,j-i,j-i)\Big\}^{1/2}$$

$$= \delta^*\ \text{(say)},$$

and that equality holds if and only if

$$F^{-1}(u)\, F^{-1}(v) = \pm \frac{1}{\delta^*}\left\{\frac{g^*(u,v) - g^*(u,1-v)}{2}\right\},$$

where

$$g^*(u,v) = \frac{1}{4B(i,j-i,n-j+1)}\left\{u^{i-1}(1-v)^{n-j} + u^{n-j}(1-v)^{i-1}\right\}(v-u)^{j-i-1},\quad 0 < u < v < 1$$

$$= \frac{1}{4B(i,j-i,n-j+1)} \left\{(1-u)^{i-1}v^{n-j}+(1-u)^{n-j}v^{i-1}\right\} (u-v)^{j-i-1}, \quad 0 < v < u < 1.$$

11. (Sugiura (1964)). Let $\{\phi_k(u)\}_{k=0}^{\infty}$, with $\phi_0 \equiv 1$, be a sequence of complete orthonormal functions in $L^2(0,1)$. Then for an arbitrary symmetric distribution with mean 0, variance 1 and $E(X^4)$ finite, show that for $1 \le i < j \le n$ and $t \ge 1$

$$\left|\mu_{i,j:n}^{(2,2)} - \left[\mu_{i:n}^{(2)} + \mu_{j:n}^{(2)}\right] + 1 - \frac{1}{2} \sum_{k,\ell=1}^{t} a_{2k}^{*} a_{2\ell}^{*} \left[b_{2k,2\ell}+b_{2\ell,2k}\right]\right|$$

$$\le \left\{[E(X^4)-1]^2 - \sum_{k,\ell=1}^{t} a_{2k}^{*2}\, a_{2\ell}^{*2}\right\}^{1/2} \left\{K_1 + K_2 - K_3 + 1 - \frac{1}{4} \sum_{k,\ell=1}^{t} \left[b_{2k,2\ell} + b_{2\ell,2k}\right]^2\right\}^{1/2},$$

where

$$a_k^{*} = \int_0^1 \{F^{-1}(u)\}^2\, \phi_k(u)\, du,$$

$b_{k,\ell}$ is as defined in Exercise 8, K_1 and K_2 are as defined in Exercise 10, and

$$K_3 = \frac{B(2i-1,2n-2i+1) + B(n,n)}{4[B(i,n-i+1)]^2} + \frac{B(2j-1,2n-2j+1) + B(n,n)}{4[B(j,n-j+1)]^2} + \frac{B(i+j-1,2n-i-j+1) + B(n+i-j,n-i+j)}{2B(i,n-i+1)\, B(j,n-j+1)};$$

for $1 \le i < j \le n$ and $t \ge 0$,

$$\left|\mu_{i,j:n}^{(2,1)} - \mu_{j:n} - \sum_{k,\ell=0}^{t} a_{2k+2}^{*}\, a_{2\ell+1}\, b_{2k+2,2\ell+1}\right|$$

$$\le \left\{E(X^4) - 1 - \sum_{k,\ell=0}^{t} a_{2k+2}^{*2}\, a_{2\ell+1}^{2}\right\}^{1/2} \left\{K_4 - K_5 - \sum_{k,\ell=0}^{t} b_{2k+2,2\ell+1}^{2}\right\}^{1/2},$$

where a_k and $b_{k,\ell}$ are as defined in Exercise 8,

$$K_4 = \frac{B(2i-1,2j-2i-1,2n-2j+1)}{4[B(i,j-i,n-j+1)]^2} - \frac{B(2j-1,2n-2j+1) - B(n,n)}{2[B(j,n-j+1)]^2}$$

and

$$K_5 = \frac{I(2i-1,n-j+1,n-j+1,j-i,j-i) - I(2n-2j+1,i,i,j-i,j-i)}{4[B(i,j-i,n-j+1)]^2};$$

for $1 \le i < j \le n$ and $t \ge 0$,

$$\left|\mu_{i,j:n}^{(1,2)} - \mu_{i:n} - \sum_{k,\ell=0}^{t} a_{2k+1}\, a_{2\ell+2}^{*}\, b_{2k+1,2\ell+2}\right|$$

$$\le \left\{E(X^4) - 1 - \sum_{k,\ell=0}^{t} a_{2k+1}^{2}\, a_{2\ell+2}^{*2}\right\}^{1/2} \left\{K_6 + K_5 - \sum_{k,\ell=0}^{t} b_{2k+1,2\ell+2}^{2}\right\}^{1/2},$$

where

$$K_6 = \frac{B(2i-1,2j-2i-1,2n-2j+1)}{4[B(i,j-i,n-j+1]^2} - \frac{B(2i-1,2n-2j+1) - B(n,n)}{2[B(i,n-i+1)]^2}.$$

12. (David and Johnson (1954)). (i) Following the notations of Section 2, for odd values of n and m = (n+1)/2 show that

$$E(X_{m:n}) \simeq G_m + \frac{1}{8(n+2)} G''_m + \frac{1}{128(n+2)^2} G^{iv}_m,$$

$$E(X_{m:n} - \mu_{m:n})^2 \simeq \frac{1}{4(n+2)} (G'_m)^2 + \frac{1}{32(n+2)^2} \left\{2G''_m G'''_m + (G''_m)^2\right\},$$

$$E(X_{m:n} - \mu_{m:n})^3 \simeq \frac{3}{16(n+2)^2} (G'_m)^2 G''_m,$$

$$E(X_{m:n} - \mu_{m:n})^4 - 3\{E(X_{m:n} - \mu_{m:n})^2\}^2$$

$$\simeq \frac{1}{16(n+2)^3} \left\{3(G'_m G''_m)^2 + (G'_m)^3 G'''_m - 6(G'_m)^4\right\},$$

$$\beta_1(X_{m:n}) = \frac{E(X_{m:n} - \mu_{m:n})^3}{\{E(X_{m:n} - \mu_{m:n})^2\}^{3/2}} \simeq \frac{3}{2\sqrt{n+2}} \frac{G''_m}{G'_m},$$

and

$$\beta_2(X_{m:n}) = \frac{E(X_{m:n} - \mu_{m:n})^4 - 3\{E(X_{m:n} - \mu_{m:n})^2\}^2}{\{E(X_{m:n} - \mu_{m:n})^2\}^2}$$

$$\simeq \frac{1}{n+2} \left\{\frac{G'''_m}{G'_m} + 3\left[\frac{G''_m}{G'_m}\right]^2 - 6\right\}.$$

(ii) If $x'_{m,n,\alpha}$ is the standardized upper 100α% point of the distribution of the median, that is,

$$\Pr\left[\frac{X_{m:n} - \mu_{m:n}}{\left\{E\left[X_{m:n} - \mu_{m:n}\right]^2\right\}^{1/2}} > x'_{m,n,\alpha}\right] = \alpha,$$

then by using the inverse Edgeworth expansion show that

$$x'_{m,n,\alpha} \simeq z_{(\alpha)} + \frac{1}{4\sqrt{n+2}} \frac{G''_m}{G'_m} (z^2_{(\alpha)} - 1)$$

$$+ \frac{1}{48(n+2)}\left\{2\left[\frac{G'''_m}{G'_m} - 6\right] \left[z^3_{(\alpha)} - 3z_{(\alpha)}\right] - 3\left[\frac{G''_m}{G'_m}\right]^2 z_{(\alpha)}\right\},$$

where $z_{(\alpha)}$ is the upper 100α% point of the standard normal distribution.

13. (David and Johnson (1954)). For the Uniform (0,1) distribution, denoting $\bar{U}_k$ for $\frac{1}{k}\sum_{i=1}^{k} (U_{i:n} - \frac{i}{n+1})$

show that the moments of $\bar{U}_k$ are given by the following equations:

$$E(\bar{U}_k) = 0,$$

$$E(\bar{U}^2_k) = \frac{k+1}{12k(n+1)^2(n+2)} \{2(n+1)(2k+1) - 3k(k+1)\},$$

$$E(\bar{U}^3_k) = \frac{(k+1)^2(n-k+1)(n-k)}{2k(n+1)^3(n+2)(n+3)}$$

and

$$E(\bar{U}^4_k) - 3\{E(\bar{U}^2_k)\}^2 = \frac{k+1}{120k^3(n+1)^4(n+2)(n+3)(n+4)} \Big\{24(n+1)^3 (2k+1)(3k^2+3k-1)$$

$$- 40(n+1)^2 k(k+1) (13k^2+13k+1) + 300 (n+1) k^2(k+1)^2(2k+1)$$

$$- 225 k^3 (k+1)^3 - 5 \frac{k(k+1)}{n+2} [2(n+1) (2k+1) - 3k(k+1)]^2\Big\}.$$

14. (Mathai (1975)). Let $\phi_o \equiv 1, \phi_1, \phi_2, \ldots$ be an orthonormal system in $L^2(0,1)$. Then for an arbitrary standardized distribution with $\mu = 0$ and $\sigma = 1$ and for $1 \le i_1 < i_2 < \ldots < i_k \le n$, by denoting

$$B = B(i_1, i_2 - i_1, \ldots, i_k - i_{k-1}, n - i_k + 1) = \frac{\Gamma(i_1)\,\Gamma(i_2 - i_1)\ldots\Gamma(n - i_k + 1)}{\Gamma(n+1)},$$

$$a_\lambda = \int_0^1 F^{-1}(u)\,\phi_\lambda(u)\,du$$

and

$$b_{\lambda_1,\ldots,\lambda_k} = B^{-1} \int\limits_{0<u_1<\ldots<u_k<1}\!\!\int u_1^{i_1-1}(u_2-u_1)^{i_2-i_1-1}(u_3-u_2)^{i_3-i_2-1}\ldots (1-u_k)^{n-i_k}\,\phi_{\lambda_1}(u_1)\ldots\phi_{\lambda_k}(u_k)\,du_1\ldots du_k,$$

show that for $t \ge 1$

$$\left|E(X_{i_1:n}X_{i_2:n}\ldots X_{i_k:n}) - \sum_{\lambda_1=1}^{t}\cdots\sum_{\lambda_k=1}^{t} a_{\lambda_1}\ldots a_{\lambda_k}\left\{\frac{1}{k!}\sum_{(r_1,\ldots,r_k)} b_{\lambda_{r_1},\ldots,\lambda_{r_k}}\right\}\right|$$

$$\le \left\{1 - \sum_{\lambda_1=1}^{t}\cdots\sum_{\lambda_k=1}^{t} a_{\lambda_1}^2\ldots a_{\lambda_k}^2\right\}^{1/2} \left[\frac{1}{k!B^2} B(2i_1-1,\, 2i_2-2i_1-1, \ldots, 2i_k-2i_{k-1}-1,\, 2n-2i_k+1)\right.$$

$$+(-1)^k + \frac{(-1)^{k-1}}{k}\sum_{r_1=1}^{k}\sum_{s_1=1}^{k} \frac{B(i_{r_1}+i_{s_1}-1, 2n-i_{r_1}-i_{s_1}+1)}{B(i_{r_1}, n-i_{r_1}+1)\,B(i_{s_1}, n-i_{s_1}+1)}$$

$$+\frac{(-1)^{k-2}}{k(k-1)}\sum_{\substack{r_1=1\\ r_1<r_2}}^{k}\sum_{r_2=1}^{k}\sum_{\substack{s_1=1\\ s_1<s_2}}^{k}\sum_{s_2=1}^{k} \frac{B(i_{r_1}+i_{s_1}-1,\, i_{r_2}-i_{r_1}+i_{s_2}-i_{s_1}-1, 2n-i_{r_2}-i_{s_2}+1)}{B(i_{r_1}, i_{r_2}-i_{r_1}, n-i_{r_2}+1)\,B(i_{s_1}, i_{s_2}-i_{s_1}, n-i_{s_2}+1)}$$

$$+\ldots - \frac{1}{k!}\sum_{r_1=1}^{k}\cdots\sum_{r_{k-1}=1}^{k}\ \sum_{s_1=1}^{k}\cdots\sum_{s_{k-1}=1}^{k}$$

$$r_1 < \ldots < r_{k-1} \qquad s_1 < \ldots < s_{k-1}$$

$$\frac{B(i_{r_1}+i_{s_1}-1, i_{r_2}-i_{r_1}+i_{s_2}-i_{s_1}-1, \ldots, 2n-i_{r_{k-1}}-i_{s_{k-1}}+1)}{B(i_{r_1}, i_{r_2}-i_{r_1}, \ldots, i_{r_{k-1}}-i_{r_{k-2}}, n-i_{r_{k-1}}+1)\,B(i_{s_1}, i_{s_2}-i_{s_1}, \ldots, n-i_{s_{k-1}}+1)}$$

$$\left.-\sum_{\lambda_1=1}^{t}\cdots\sum_{\lambda_k=1}^{t}\left\{\frac{1}{k!}\sum_{(r_1,\ldots,r_k)} b_{\lambda_{r_1},\ldots,\lambda_{r_k}}\right\}^2\right]^{1/2},$$

where $\sum_{(r_1,\ldots,r_k)}$ denotes the sum over all permutations of the integers $1,2,\ldots,k$.

15. (Mathai (1976)). Let the parent distribution be symmetric about zero. Let $\phi_o \equiv 1, \phi_1, \phi_2, \ldots$ be a

complete orthonormal system in $L^2(0,1)$ such that $\phi_j(u) = (-1)^j \phi_j(1-u)$, $j = 1,2,\ldots$ Then for $1 \le i_1 < i_2 < \ldots < i_k \le n$, show that for $t \ge 0$

$$\left| E(X_{i_1:n} X_{i_2:n} \cdots X_{i_k:n}) - \sum_{\lambda_1=0}^{t} \cdots \sum_{\lambda_k=0}^{t} a_{2\lambda_1+1} \cdots a_{2\lambda_k+1} \right.$$

$$\left. \left\{ \frac{1}{k!} \sum_{(r_1,\ldots,r_k)} b_{2\lambda_{r_1}+1,\ldots,2\lambda_{r_k}+1} \right\} \right| \le \left\{ 1 - \sum_{\lambda_1=0}^{t} \cdots \sum_{\lambda_k=0}^{t} a^2_{2\lambda_1+1} \cdots a^2_{2\lambda_k+1} \right\}^{1/2}$$

$$\times \left\{ \sum_{\lambda_1=0}^{\infty} \cdots \sum_{\lambda_k=0}^{\infty} b^{*2}_{2\lambda_1+1,\ldots,2\lambda_k+1} - \sum_{\lambda_1=0}^{t} \cdots \sum_{\lambda_k=0}^{t} b^{*2}_{2\lambda_1+1,\ldots,2\lambda_k+1} \right\}^{1/2},$$

where

$$b^{*2}_{2\lambda_1+1,\ldots,2\lambda_k+1} = \left\{ \frac{1}{k!} \sum_{(r_1,\ldots,r_k)} b_{2\lambda_{r_1}+1,\ldots,2\lambda_{r_k}+1} \right\}^2$$

and, as before, $\sum_{(r_1,\ldots,r_k)}$ denotes the sum over all permutations of the integers $1,2,\ldots,k$.

16. (Chu and Hotelling (1955); Siddiqui (1962)). (i) Let X be a random variable with cdf $F(x)$, pdf $f(x)$, and a unique median which may be taken as zero without loss of any generality. Suppose that $m(z)$, the inverse function of $z(m) = 2F(m)-1$, is for $-1 < z < 1$ uniquely defined and equal to a convergent series of powers of z; let

$$m(z) = a_1 z \left[1 + \sum_{p=1}^{\infty} c_p z^p \right].$$

Then, show that for $k = 0,1,2,\ldots$

$$E\left[X^{2k}_{n+1:2n+1}\right]$$

$$= \frac{a_1^{2k} B(n+1, k+\frac{1}{2})}{B(n+1, \frac{1}{2})} \left[1 + \frac{2k+1}{2n+2k+3} b_2(k) + \frac{(2k+1)(2k+3)}{(2n+2k+3)(2n+2k+5)} b_4(k) + \ldots \right]$$

and

$$E(X^{2k+1}_{n+1:2n+1}) = \frac{(2k+1) a_1^{2k+1} B(n+1, k+\frac{3}{2})}{B(n+1, \frac{1}{2})} \left[c_1 + \frac{2k+3}{2n+2k+5} b_3(k) \right.$$

$$\left. + \frac{(2k+3)(2k+5)}{(2n+2k+5)(2n+2k+7)} b_5(k) + \ldots \right],$$

where $b_2(k), b_3(k), b_4(k), \ldots$ are expressible in terms of $k, c_1, c_2, \ldots$; for example,

$$b_2(k) = k(2k-1)c_1^2 + 2kc_2,$$

$$b_3(k) = \frac{1}{3} k\,(2k-1)c_1^3 + 2kc_1c_2 + c_3,$$

etc.,

(ii) If $c_p/p^k \to$ constant (which may be zero) for some integer k, then the series in (i) are convergent for all integers $n \ge n_o$.

(iii) If f(x) is symmetrical about 0, then show that $c_1 = c_3 = \ldots = 0$ and also that $E(X_{n+1:2n+1}^{2k+1}) = 0$, whenever it exists, for k = 0,1,2,...

(iv) Let F_1 and F_2 be two different symmetric populations. Let $m_1(z)$ and $m_2(z)$ correspond to F_1 and F_2, respectively. Further, let $X_{n+1:2n+1}$ and $Y_{n+1:2n+1}$ be the medians in samples of size 2n+1 drawn from the populations F_1 and F_2, respectively. If $m_1(z) \le m_2(z)$ for $0 < z < 1$, where strict inequality holds over an interval of non-zero length, then show that for k = 1,2,...

$$E\left[X_{n+1:2n+1}^{2k}\right] \le E\left[Y_{n+1:2n+1}^{2k}\right].$$

17. (Tiku and Malik (1972)). Let $X_{i:n}$ $(1 \le i \le n)$ be the i'th order statistic in a sample of size n from an arbitrary distribution. Using the three-moment chi-square- and t-approximations given by

$$\chi_f^2 \simeq (X_{i:n}+a)/g \text{ and } t_v = (X_{i:n}+d)/h,$$

where

$$f = 8/\beta_1 \text{ (the degree of freedom of } \chi^2\text{)}, \; g = \tfrac{1}{4}\,\mu_3/\mu_2 \text{ and } a = gf - \mu_1' \; (\mu_3 > 0)$$

and

$$v = 4\,(\beta_2-1.5)/(\beta_2-3) \text{ (the degree of freedom of t)},$$

$$h = \sqrt{\mu_2(1-\tfrac{2}{v})} \text{ and } d = -\mu_1',$$

with μ_1' as the mean, μ_k as the k'th central moment of $X_{i:n}$, $\beta_1 = \mu_3^2/\mu_2^3$ and $\beta_2 = \mu_4/\mu_2^2$, verify by computations that the t-approximation provides reasonably good approximations for the percentage points and the probability integrals of $X_{i:n}$ for small values of β_1, say < 0.01; similarly, if $\beta_1 \ge 0.01$, verify that the chi-square approximation provides reasonably good approximations as long as β_2 does not deviate from the β_2- value of the χ^2, viz., $3 + 1.5\beta_1$. Note that if $\mu_3 < 0$, we may replace χ_f^2 by $1 - \chi_f^2$.

Remark: You may do the comparisons for the normal, Student's t, logistic, gamma, Weibull and Pareto distributions. It should be pointed out here that these approximations may not be satisfactory for the extreme of the lower tail.

18. (Tadikamalla (1977)). (i) Let $X_{i:n}$ $(1 \le i \le n)$ be the order statistics obtained from a random sample of size n from the four parameter Burr distribution with c.d.f.

$$F(x) = 1 - \left\{1 + \left[\frac{x-a}{b}\right]^c\right\}^{-d}, \quad x \ge a,$$

where d,c > 0 are the shape parameters, b > 0 is the scale parameter and a is the location parameter. Then show that for $k \ge 1$

$$E(X_{i:n}-a)^k = \frac{n!}{(i-1)!\,(n-i)!}\, b^k\, d \sum_{j=0}^{i-1} (-1)^j \frac{(i-1)!}{j!(i-1-j)!}\, B(nd - id + d - \tfrac{k}{c} + jd,\, 1 + \tfrac{k}{c}),$$

where B(.,.) is the complete beta function.

(ii) By determining the parameters d,c,b and a of the Burr distribution that approximates the gamma distribution with pdf

$$f^*(x) = e^{-x}\, x^{\alpha-1}/\Gamma(\alpha), \; x \ge 0, \; \alpha > 0,$$

for any given α by matching their first four moments (mean, variance, skewness and kurtosis), compute the approximate values of the first four moments of all order statistics from the above gamma distribution for sample sizes up to 10. By comparing them with the exact values given by Gupta (1960), comment on the accuracy of this approximation procedure.

CHAPTER 5

ORDER STATISTICS FROM A SAMPLE CONTAINING A SINGLE OUTLIER

5.0. Introduction

Density functions and joint density functions of order statistics arising from a sample containing a single outlier have been given by Shu (1978) and David and Shu (1978), and have been made use of by David et al. (1977) in tabulating means, variances and covariances of order statistics from a normal sample comprising one outlier. One may also refer to Vaughan and Venables (1972) for more general expressions of distributions of order statistics when the sample, in fact, includes k outliers. The importance of a systematic study of the order statistics from an outlier model and the usefulness of the tables of means, variances and covariances of these order statistics in the context of robustness has been well demonstrated by several authors including Andrews et al. (1972), David and Shu (1978), David (1981), and Tiku et al. (1986).

In this Chapter, we first derive several recurrence relations and identities satisfied by the single and the product moments of order statistics. These relations are then applied in order to determine the maximum number of single and double integrals to be evaluated for the calculation of the first two single moments and the product moments of order statistics in a sample of size n, assuming these quantities for all sample sizes less than n to be known. All these results are presented in Sections 2 through 4. Results of similar nature for the symmetric outlier case, i.e., the case when all the n observations including the outlying observation have distributions symmetric about zero, are presented in Section 5. In Section 6, we derive some relations between the moments of order statistics from a symmetric outlier model and the moments of order statistics from a single outlier model obtained by folding both the distributions at zero and also discuss the cumulative rounding error committed by using these recurrence relations. These results have all been established recently in a series of papers by Balakrishnan (1987b, 1988a,b,c) and they generalize the results presented in Chapter 2. The functional behaviour of order statistics from location- and scale-outlier models is discussed in detail in Section 7. Finally, in Section 8 we make use of these results to study the bias and mean square error of various omnibus robust estimators of the location parameter in the normal case. Work of this nature has also been carried out earlier by Crow and Siddiqui (1967), Gastwirth and Cohen (1970), Andrews et al. (1972), and Tiku (1980). By considering a single outlier exponential model, Kale and Sinha (1971) and Joshi (1972) have made investigations regarding the mean square error of a class of estimators of the mean of the exponential distribution; see also Barnett and Lewis (1978) and Lawless (1982) for further details on this topic.

Throughout this Chapter results are derived under the assumption that the distributions under discussion are absolutely continuous. Since any distribution function can be represented as a weak limit of a sequence of absolutely continuous distributions, all results which do not explicitly refer to densities continue to hold for general (not necessarily absolutely continuous) distributions. See, for example, Exercises 17–19.

5.1. Distributions of order statistics

We shall derive here the distributions of order statistics obtained from a sample of size n when an unidentified single outlier is present in the sample. For convenience, let us represent the sample by n independent absolutely continuous random variables X_r ($r = 1,2,\ldots,n-1$) and Y, such that X_r has pdf f(x) and cdf F(x), and Y has pdf g(x) and cdf G(x). Further, let

$$Z_{1:n} \le Z_{2:n} \le \ldots \le Z_{n:n} \tag{5.1}$$

be the order statistics obtained by arranging the n independent observations in increasing order of magnitude.

The cdf of $Z_{n:n}$, denoted by $h_{n:n}(x)$, may now be obtained as

$$\begin{aligned} H_{n:n}(x) &= \Pr\{\text{all of } X_1,X_2,\ldots,X_{n-1},\ Y \le x\} \\ &= \{F(x)\}^{n-1}\, G(x). \end{aligned} \tag{5.2}$$

Similarly, the cdf of $Z_{i:n}$ $(1 \le i \le n-1)$ may be obtained as

$$\begin{aligned} H_{i:n}(x) &= \Pr\{\text{at least } i \text{ of } X_1,X_2,\ldots,X_{n-1},\ Y \le x\} \\ &= \Pr\{\text{exactly } i-1 \text{ of } X_1,X_2,\ldots,X_{n-1} \le x \text{ and } Y \le x\} \\ &\qquad + \Pr\{\text{at least } i \text{ of } X_1,X_2,\ldots,X_{n-1} \le x\} \\ &= \binom{n-1}{i-1} \{F(x)\}^{i-1} \{1-F(x)\}^{n-i}\, G(x) + F_{i:n-1}(x), \end{aligned} \tag{5.3}$$

where $F_{i:n-1}(x)$ is the cdf of the i'th order statistic in a sample of size n–1 drawn from a population with pdf f(x) and cdf F(x). Differentiating the expressions in (5.2) and (5.3), we obtain the density function of $Z_{i:n}$ $(1 \le i \le n)$ as

$$\begin{aligned} h_{i:n}(x) &= \frac{(n-1)!}{(i-2)!(n-i)!} \{F(x)\}^{i-2} \{1-F(x)\}^{n-i}\, G(x)\, f(x) \\ &+ \frac{(n-1)!}{(i-1)!(n-i)!} \{F(x)\}^{i-1} \{1-F(x)\}^{n-i}\, g(x) \\ &+ \frac{(n-1)!}{(i-1)!(n-i-1)!} \{F(x)\}^{i-1} \{1-F(x)\}^{n-i-1} \{1-G(x)\}\, f(x), \quad -\infty < x < \infty, \end{aligned} \tag{5.4}$$

where the first term drops out if $i = 1$, and the last term if $i = n$.

In view of the importance of this result we will also give an alternative proof which lends itself to further extensions. The event $x < Z_{i:n} \le x+\delta x$ may be seen as follows

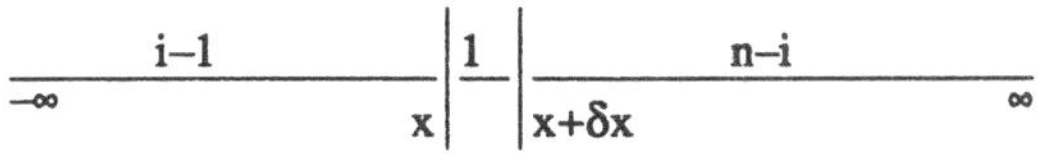

$Z_r \le x$ for i–1 of the Z_r, $x < Z_r \le x+\delta x$ for exactly one Z_r, and $Z_r > x+\delta x$ for the remaining n–i of the Z_r. Realizing now that the outlying observation Y could belong to any one of three parcels, we have the following three possibilities:

(i) $X_r \le x$ for i–2 of the X_r and $Y \le x$, $x < X_r \le x+\delta x$ for exactly one X_r, and $X_r > x+\delta x$ for the remaining n–i of the X_r, with a probability

$$\frac{(n-1)!}{(i-2)!1!(n-i)!} \{F(x)\}^{i-2}\, G(x)\{F(x+\delta x) - F(x)\}\{1-F(x+\delta x)\}^{n-i};$$

(ii) $X_r \le x$ for i–1 of the X_r, $x < Y \le x+\delta x$, and $X_r > x+\delta x$ for the remaining n–i of the X_r, with a probability

$$\frac{(n-1)!}{(i-1)!0!(n-i)!} \{F(x)\}^{i-1} \{G(x+\delta x) - G(x)\}\{1-F(x+\delta x)\}^{n-i};$$

and

(iii) $X_r \leq x$ for $i-1$ of the X_r, $x < X_r \leq x+\delta x$ for exactly one X_r, and $X_r > x+\delta x$ for the remaining $n-i-1$ of the X_r and $Y > x+\delta x$, with a probability

$$\frac{(n-1)!}{(i-1)!1!(n-i-1)!} \{F(x)\}^{i-1} \{F(x+\delta x) - F(x)\}\{1-F(x+\delta x)\}^{n-i-1}\{1-G(x+\delta x)\}.$$

Regarding δx as small, we could write

$$\Pr(x < Z_{i:n} \leq x+\delta x)$$

$$\equiv \frac{(n-1)!}{(i-2)!(n-i)!} \{F(x)\}^{i-2} G(x) \{1-F(x+\delta x)\}^{n-i} f(x)\, \delta x$$

$$+ \frac{(n-1)!}{(i-1)!(n-i)!} \{F(x)\}^{i-1} \{1-F(x+\delta x)\}^{n-i} g(x)\, \delta x$$

$$+ \frac{(n-1)!}{(i-1)!(n-i-1)!} \{F(x)\}^{i-1} \{1-F(x+\delta x)\}^{n-i-1} \{1-G(x+\delta x)\} f(x)\, \delta x + 0\left[(\delta x)^2\right] \tag{5.5}$$

where $0\left[(\delta x)^2\right]$ denotes the probability of more than one Z_r falling in the interval $(x,x+\delta x)$ and hence is a term of order $(\delta x)^2$. Dividing both sides of (5.5) by δx and letting $\delta x \to 0$, we once again obtain the density function of $Z_{i:n}$ as in equation (5.4).

Proceeding exactly on similar lines by noting that the joint event $x < Z_{i:n} \leq x+\delta x$, $y < Z_{j:n} \leq y + \delta y$ is obtained by the configuration (neglecting terms of lower order of probability)

$$\underset{-\infty}{\underline{\quad i-1 \quad}}\left|\underset{x}{\underline{1}}\right|\underset{x+\delta x}{\underline{\quad j-i-1 \quad}}\left|\underset{y}{\underline{1}}\right|\underset{y+\delta y}{\underline{\quad n-j \quad}}_{\infty}$$

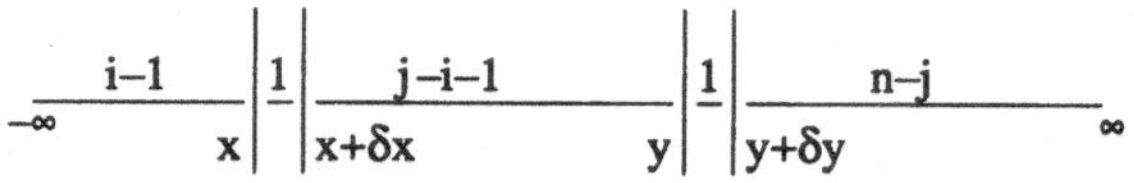

we obtain the joint density function of $Z_{i:n}$ and $Z_{j:n}$ $(1 \leq i < j \leq n)$ as

$$h_{i,j:n}(x,y) = \frac{(n-1)!}{(i-2)!(j-i-1)!(n-j)!} \{F(x)\}^{i-2} \{F(y)-F(x)\}^{j-i-1}\{1-F(y)\}^{n-j} G(x)\, f(x)\, f(y)$$

$$+ \frac{(n-1)!}{(i-1)!(j-i-1)!(n-j)!} \{F(x)\}^{i-1} \{F(y)-F(x)\}^{j-i-1}\{1-F(y)\}^{n-j} g(x)\, f(y)$$

$$+ \frac{(n-1)!}{(i-1)!(j-i-2)!(n-j)!} \{F(x)\}^{i-1} \{F(y)-F(x)\}^{j-i-2}\{1-F(y)\}^{n-j} \{G(y)-G(x)\}\, f(x)f(y)$$

$$+ \frac{(n-1)!}{(i-1)!(j-i-1)!(n-j)!} \{F(x)\}^{i-1}\{F(y)-F(x)\}^{j-i-1}\{1-F(y)\}^{n-j} f(x)\, g(y)$$

$$+ \frac{(n-1)!}{(i-1)!(j-i-1)!(n-j-1)!} \{F(x)\}^{i-1}\{F(y)-F(x)\}^{j-i-1}\{1-F(y)\}^{n-j-1}\{1-G(y)\}f(x)\, f(y),$$

$$-\infty < x < y < \infty, \tag{5.6}$$

where the first term drops out if $i = 1$, the last term if $j = n$, and the middle term if $j = i+1$. Note that the densities in equations (5.4) and (5.6) are special cases of a very general result of Vaughan and Venables (1972). They have essentially expressed the joint density function of k order statistics arising from n absolutely continuous populations in the form of a permanent.

5.2. Relations for single moments

With the density function of $Z_{i:n}$ as in equation (5.4), we have the single moment $\mu_{i:n}^{(k)}$ $(1 \leq i \leq n,$ $k \geq 1)$ as

$$\mu_{i:n}^{(k)} = \int_{-\infty}^{\infty} x^k\, h_{i:n}(x)\, dx. \tag{5.7}$$

Further, let $f_{i:n}(x)$ be the pdf of the i'th order statistic in a sample of size n drawn from a continuous population with pdf f(x) and cdf F(x), and the single moments of order statistics for this case be denoted by

$$\nu_{i:n}^{(k)} = \int_{-\infty}^{\infty} x^k f_{i:n}(x)\, dx$$

$$= \{B(i,n-i+1)\}^{-1} \int_{-\infty}^{\infty} x^k \{F(x)\}^{i-1} \{1-F(x)\}^{n-i} f(x), \tag{5.8}$$

for $1 \le i \le n$ and $k = 1,2,\ldots$. Then these single moments of order statistics, viz., $\mu_{i:n}^{(k)}$ and $\nu_{i:n}^{(k)}$, satisfy the following recurrence relations and identities for arbitrary continuous distributions F and G. As mentioned by David et al. (1977), these results also provide useful checks in assessing the accuracy of the computation of moments of order statistics from a single outlier model.

Relation 5.1: For $n \ge 2$ and $k = 1,2,\ldots$,

$$\sum_{i=1}^{n} \mu_{i:n}^{(k)} = (n-1)\, \nu_{1:1}^{(k)} + \mu_{1:1}^{(k)}. \tag{5.9}$$

Proof. By considering the identity

$$\sum_{i=1}^{n} Z_{i:n}^{k} = \sum_{i=1}^{n-1} X_i^k + Y^k$$

and taking expectations on both sides, we immediately obtain the required relation.

Setting $k = 1$ and 2, in particular, we derive the results

$$\sum_{i=1}^{n} \mu_{i:n} = (n-1)\, \nu_{1:1} + \mu_{1:1}$$

and

$$\sum_{i=1}^{n} \mu_{i:n}^{(2)} = (n-1)\, \nu_{1:1}^{(2)} + \mu_{1:1}^{(2)},$$

which have been made use of by David et al. (1977).

Relation 5.2: For $1 \le i \le n-1$ and $k = 1,2,\ldots$,

$$i\, \mu_{i+1:n}^{(k)} + (n-i)\, \mu_{i:n}^{(k)} = (n-1)\, \mu_{i:n-1}^{(k)} + \nu_{i:n-1}^{(k)}. \tag{5.10}$$

Proof. First, consider the expression of $i\, \mu_{i+1:n}^{(k)}$ from equations (5.4) and (5.7) and split the first term into two by writing the multiple i as $(i-1) + 1$. Next, consider the expression of $(n-i)\, \mu_{i:n}^{(k)}$ and split the last term into two by writing the multiple $n-i$ as $(n-i-1) + 1$. Finally, adding the two expressions and simplifying the resulting equation, we obtain the relation (5.10).

Relation 5.2 has been derived by Balakrishnan (1988a) and it is easy to see that we just require the value of the k'th moment of a single order statistic in a sample of size n as Relation 5.2 would enable us to compute the k'th moment of the remaining $n-1$ order statistics, using of course the moments $\mu^{(k)}$ and $\nu^{(k)}$ in samples of size less than n. In particular, setting $n = 2m$ and $i = m$ in equation (5.10) we obtain the following relation.

Relation 5.3. For even values of n, say $n = 2m$, and $k \ge 1$,

$$\frac{1}{2}\left\{\mu_{m+1:2m}^{(k)} + \mu_{m:2m}^{(k)}\right\} = \left[1 - \frac{1}{2m}\right] \mu_{m:2m-1}^{(k)} + \frac{1}{2m}\, \nu_{m:2m-1}^{(k)}$$

$$= \mu^{(k)}_{m:2m-1} + \frac{1}{2m}\left[\nu^{(k)}_{m:2m-1} - \mu^{(k)}_{m:2m-1}\right]. \tag{5.11}$$

For k = 1, in particular, we have the expected value of the median in a sample of even size (n = 2m) on the LHS of (5.11), while on the RHS we have the expected value of the median in a sample of odd size (n = 2m–1) and an additive biasing factor involving the difference of the expected values of the medians in a sample of size 2m–1 from the outlier and the non–outlier models.

Relation 5.4: For $1 \le i \le n-1$ and k = 1,2,...,

$$\mu^{(k)}_{i:n} = \sum_{j=i}^{n} (-1)^{j-i} \binom{n-1}{j-1}\binom{j-1}{i-1} \mu^{(k)}_{j:j} + \nu^{(k)}_{i:n-1}. \tag{5.12}$$

Proof. This relation is simply obtained by considering the expression of $\mu^{(k)}_{i:n}$ from equations (5.4) and (5.7), upon expanding the term $\{1-F(x)\}^m$, (m = n–i, n–i–1), binomially in powers of F(x), and finally simplifying the resulting expression by making use of equation (5.7) and Relation 2.6.

Relation 5.5: For $2 \le i \le n$ and k = 1,2,...,

$$\mu^{(k)}_{i:n} = \sum_{j=n-i+1}^{n} (-1)^{j-n+i-1} \binom{n-1}{j-1}\binom{j-1}{n-i} \mu^{(k)}_{1:j} + \nu^{(k)}_{i-1:n-1}. \tag{5.13}$$

Proof. First, consider the expression of $\mu^{(k)}_{i:n}$ from equations (5.4) and (5.7). Next, write the term $\{F(x)\}^m$ as $[1-\{1-F(x)\}]^m$, (m = i–2,i–1) and expand it binomially in powers of {1–F(x)}. Relation (5.13) now follows immediately upon simplifying the resulting expression by making use of equation (5.7) and Relation 2.8.

Remark 5.6: Relations 5.4 and 5.5 have both been derived by Balakrishnan (1988a) and they are quite important as they usefully express the k'th moment of the i'th order statistic from a sample of size n in terms of the k'th moment of the largest and the smallest order statistics in samples of size n and less, respectively. In addition, they also involve the k'th moment of the i'th and (i–1)'th order statistics in a sample of size n–1 from the non–outlier model. In any case, we could note once again from these two relations that we just require the value of the k'th moment of a single order statistic (either the smallest or the largest) in a sample of size n in order to compute the k'th moment of the remaining n–1 order statistics, given the moments $\mu^{(k)}$ and $\nu^{(k)}$ in samples of size less than n. Note that this conforms with the comment made earlier based on Relation 5.2. This is quite expected as both Relations 5.4 and 5.5 could be derived by repeated application of Relation 5.2. However, we need to be careful in employing these two recurrence relations in the computational algorithm as increasing values of n result in large combinatorial terms and hence in an error of large magnitude.

5.3. Relations for product moments

With the joint density function of $Z_{i:n}$ and $Z_{j:n}$ as in equation (5.6), we have the product moment $\mu_{i,j:n}$ $(1 \le i < j \le n)$ as

$$\mu_{i,j:n} = \iint_{W_1} xy\, h_{i,j:n}(x,y)\, dy\, dx \tag{5.14}$$

$$= \iint_{W_2} xy\, h_{i,j:n}(y,x)\, dy\, dx, \tag{5.15}$$

where $W_1 = \{(x,y) : -\infty < x < y < \infty\}$ and $W_2 = \{(x,y): -\infty < y < x < \infty\}$. Further, let $f_{i,j:n}(x,y)$ be the joint pdf of the i'th and j'th order statistics in a sample of size n drawn from a continuous population with pdf f(x) and cdf F(x), and the product moments of order statistics for this case be denoted by

$$\nu_{i,j:n} = \iint_{W_1} xy\, f_{i,j:n}(x,y)\, dy\, dx \tag{5.16}$$

$$= \iint_{W_2} xy\, f_{i,j:n}(y,x)\, dy\, dx, \tag{5.17}$$

for $1 \le i < j \le n$. Then these product moments of order statistics, viz., $\mu_{i,j:n}$ and $\nu_{i,j:n}$, satisfy the following recurrence relations and identities for any arbitrary continuous distributions F and G. In addition to providing straight forward generalizations of the results given in Chapter 2 and some simple checks for assessing the accuracy of the computation of product moments, these results could also be effectively applied to reduce considerably the amount of numerical computation involved in the evaluation of means, variances and covariances of order statistics from an outlier model, at least for small sample sizes.

Relation 5.7: For $n \ge 2$,

$$\sum_{i=1}^{n} \sum_{j=1}^{n} \mu_{i,j:n} = (n-1)\nu_{1:1}^{(2)} + \mu_{1:1}^{(2)} + (n-1)(n-2)\,\nu_{1:1}^{2} + 2(n-1)\,\nu_{1:1}\mu_{1:1}. \tag{5.18}$$

Proof. The above relation follows directly by considering the identity

$$\sum_{i=1}^{n} \sum_{j=1}^{n} Z_{i:n} Z_{j:n} = \sum_{i=1}^{n} \sum_{j=1}^{n} Z_i Z_j = \sum_{i=1}^{n} Z_i^2 + \sum_{i=1}^{n} \sum_{\substack{j=1 \\ i\ne j}}^{n} Z_i Z_j$$

and taking expectation on both sides.

Relation 5.8: For $n \ge 2$,

$$\sum_{i=1}^{n-1} \sum_{j=i+1}^{n} \mu_{i,j:n} = \binom{n-1}{2} \nu_{1:1}^{2} + (n-1)\,\nu_{1:1}\mu_{1:1}. \tag{5.19}$$

Proof. This is simply derived from Relation 5.7 upon using the result that

$$\sum_{i=1}^{n} \mu_{i:n}^{(2)} = (n-1)\,\nu_{1:1}^{(2)} + \mu_{1:1}^{(2)}$$

obtained from equation (5.9).

Relations 5.7 and 5.8 are very simple to use and, hence, as pointed out by David et al. (1977), could be effectively applied to check the accuracy of the computation of the product moments.

Relation 5.9: For $2 \le i < j \le n$,

$$(i-1)\,\mu_{i,j:n} + (j-i)\,\mu_{i-1,j:n} + (n-j+1)\,\mu_{i-1,j-1:n} = (n-1)\,\mu_{i-1,j-1:n-1} + \nu_{i-1,j-1:n-1}. \tag{5.20}$$

Proof. First, consider the expression of the LHS of (5.20) obtained from equation (5.14). Now split the first term in (i–1) $\mu_{i,j:n}$ into two by writing the multiple i–1 and (i–2)+1; split the middle term in (j–i) $\mu_{i-1,j:n}$ into two by writing the multiple j–i as (j–i–1)+1; similarly, split the last term in (n–j+1) $\mu_{i-1,j-1:n}$ into two by writing the multiple n–j+1 as (n–j)+1. Finally, adding all these and simplifying the resulting expression by making use of equations (5.14) and (5.16), we obtain the RHS of (5.20).

Relation 5.9 has been derived by Balakrishnan (1988a) and it should be noted that the recurrence formula (5.20) would enable us to calculate all the product moments $\mu_{i,j:n}$ $(1 \le i < j \le n)$ by knowing n–1 suitably chosen moments, for example, $\mu_{1,2:n}, \mu_{2,3:n}, \ldots, \mu_{n-1,n:n}$.

Relation 5.10: For $1 \le i < j \le n$,

$$\mu_{i,j:n} - \nu_{i,j:n} + \sum_{r=0}^{i-1} \sum_{s=0}^{n-j} (-1)^{n-r-s} \binom{n-1}{s} \binom{n-1-s}{r} \left\{\mu_{n-j-s+1,n-i-s+1:n-r-s} - \nu_{n-j-s+1,n-i-s+1:n-r-s}\right\}$$

$$= \frac{1}{n} \sum_{r=1}^{j-i} (-1)^{j-i-r} \binom{n}{j-r} \binom{j-r-1}{i-1} \Big\{(j-r)\, \nu_{r:n-j+r}\left[\mu_{j-r:j-r} - \nu_{j-r:j-r}\right]$$

$$+ (n-j+r)\, \nu_{j-r:j-r}\left[\mu_{r:n-j+r} - \nu_{r:n-j+r}\right]\Big\}. \tag{5.21}$$

Proof. Noting that $W_1 \cup W_2 = \mathbb{R}^2$, we have for $1 \le i < j \le n$

$$\mathcal{J} = \iint_{\mathbb{R}^2} xy\, h_{i,j:n}(x,y)\, dy\, dx$$

$$= \iint_{W_1} xy\, h_{i,j:n}(x,y)\, dy\, dx + \iint_{W_2} xy\, h_{i,j:n}(x,y)\, dy\, dx$$

$$= \mu_{i,j:n} + \iint_{W_2} xy\, h_{i,j:n}(x,y)\, dy\, dx \tag{5.22}$$

upon using (5.14). By writing the term $\{F(x)\}^a$ as $[1-\{1-F(x)\}]^a$, expanding the terms $\{F(x)\}^a$ and $\{1-F(y)\}^b$ binomially in powers of 1–F(x) and F(y), respectively, in the integral over W_2, and simplifying the resulting expression using (5.15) and (5.17) we get from equation (5.22)

$$\mathcal{J} = \mu_{i,j:n} + \sum_{r=0}^{i-1} \sum_{s=0}^{n-j} (-1)^{n-r-s} \binom{n}{s} \binom{n-s}{r} \nu_{n-j-s+1,n-i-s+1:n-r-s}$$

$$+ \sum_{r=0}^{i-1} \sum_{s=0}^{n-j} (-1)^{n-r-s} \binom{n-1}{s} \binom{n-1-s}{r} \left\{\mu_{n-j-s+1,n-i-s+1:n-r-s} - \nu_{n-j-s+1,n-i-s+1:n-r-s}\right\}.$$

Making use of Relation 2.14 we may rewrite the above equation as

$$\mathcal{J} = \mu_{i,j:n} - \nu_{i,j:n} + \sum_{r=1}^{j-i} (-1)^{j-i-r} \binom{n}{j-r} \binom{j-r-1}{i-1} \nu_{j-r:j-r}\, \nu_{r:n-j+r}$$

$$+ \sum_{r=0}^{i-1} \sum_{s=0}^{n-j} (-1)^{n-r-s} \binom{n-1}{s} \binom{n-1-s}{r} \left\{\mu_{n-j-s+1,n-i-s+1:n-r-s} - \nu_{n-j-s+1,n-i-s+1:n-r-s}\right\}. \tag{5.23}$$

We could alternatively write

$$\mathcal{J} = \iint_{\mathbb{R}^2} xy\, h_{i,j:n}(x,y)\, dy\, dx$$

$$= \int_{-\infty}^{\infty} \int_{-\infty}^{\infty} xy \, h_{i,j:n}(x,y) \, dy \, dx.$$

Now expanding the term $\{F(y)-F(x)\}^a$ binomially in powers of F(x) and F(y) and simplifying the resulting expression by using equations (5.7) and (5.8), we also obtain

$$\mathcal{J} = \frac{1}{n} \sum_{r=1}^{j-i} (-1)^{j-i-r} \binom{n}{j-r} \binom{j-r-1}{i-1} \Big\{ (j-r)\, \nu_{r:n-j+r}\, \mu_{j-r:j-r} + (n-j+r)\, \nu_{j-r:j-r}\, \mu_{r:n-j+r} \Big\}. \tag{5.24}$$

Relation (5.21) follows upon equating (5.23) and (5.24).

In the above relation, established by Balakrishnan (1988a), it should be noted that only two product moments, viz., $\mu_{i,j:n}$ and $\mu_{n-j+1,n-i+1:n}$, are in samples of size n from the outlier model. In particular, for j = i+1 we have the following relation.

Relation 5.11: For i = 1,2,...,n−1,

$$\mu_{i,i+1:n} - \nu_{i,i+1:n} + (-1)^n \Big\{ \mu_{n-i,n-i+1:n} - \nu_{n-i,n-i+1:n} \Big\}$$
$$= \sum_{r=0}^{i-1} \sum_{s=1}^{n-i-1} (-1)^{n+1-r-s} \binom{n-1}{s} \binom{n-1-s}{r} \Big\{ \mu_{n-i-s,n-i-s+1:n-r-s} - \nu_{n-i-s,n-i-s+1:n-r-s} \Big\}$$
$$+ \sum_{r=1}^{i-1} (-1)^{n+1-r} \binom{n-1}{r} \Big\{ \mu_{n-i,n-i+1:n-r} - \nu_{n-i,n-i+1:n-r} \Big\}$$
$$+ \frac{1}{n} \binom{n}{i} \Big\{ i\, \nu_{1:n-i} \big[\mu_{i:i} - \nu_{i:i} \big] + (n-i)\, \nu_{i:i} \big[\mu_{1:n-i} - \nu_{1:n-i} \big] \Big\}. \tag{5.25}$$

Similarly, by setting j = n−i+1 in equation (5.21), we obtain the following relation.

Relation 5.12: For i = 1,2,...,[n/2],

$$\{1+(-1)^n\} \big[\mu_{i,n-i+1:n} - \nu_{i,n-i+1:n} \big]$$
$$= \sum_{r=0}^{i-1} \sum_{s=1}^{i-1} (-1)^{n+1-r-s} \binom{n-1}{s} \binom{n-1-s}{r} \Big\{ \mu_{i-s,n-i-s+1:n-r-s} - \nu_{i-s,n-i-s+1:n-r-s} \Big\}$$
$$+ \sum_{r=1}^{i-1} (-1)^{n+1-r} \binom{n-1}{r} \Big\{ \mu_{i,n-i+1:n-r} - \nu_{i,n-i+1:n-r} \Big\}$$
$$+ \frac{1}{n} \sum_{r=1}^{n-2i+1} (-1)^{n+1-r} \binom{n}{i+r-1} \binom{n-i-r}{i-1} \Big\{ (i+r-1)\nu_{n-i-r+1:n-i-r+1} \big[\mu_{r:i+r-1} - \nu_{r:i+r-1} \big] + (n-i-r+1)\, \nu_{r:i+r-1} \big[\mu_{n-i-r+1:n-i-r+1} - \nu_{n-i-r+1:n-i-r+1} \big] \Big\}. \tag{5.26}$$

For odd values of n, we note from Relation 5.11 that we need to calculate only (n−1)/2 product moments $\mu_{i,i+1:n}$ $(1 \le i \le \frac{n-1}{2})$. Similarly, for even values of n, Relation 5.12 shows that the product moments $\mu_{i,n-i+1:n}$ $(1 \le i \le [n/2])$ could all be obtained from the moments in samples of sizes n−1 and less. In particular, for even values of n, say n = 2m and i = 1, Relation 5.12 yields

$$2\big[\mu_{1,2m:2m} - \nu_{1,2m:2m} \big]$$

$$= \frac{1}{2m} \sum_{r=1}^{2m-1} (-1)^{r-1} \binom{2m}{r} \Big\{ r\, \nu_{2m-r:2m-r} \big[\mu_{r:r} - \nu_{r:r}\big] + (2m-r)\, \nu_{r:r} \big[\mu_{2m-r:2m-r} - \nu_{2m-r:2m-r}\big] \Big\}$$

which, upon using equation (2.26) and simplifying, yields the relation

$$\mu_{1,2m:2m} = \sum_{r=1}^{2m-1} (-1)^{r-1} \binom{2m-1}{r} \nu_{r:r}\, \mu_{2m-r:2m-r}. \tag{5.27}$$

This relation generalizes the result of Govindarajulu (1963a) to the case when the order statistics arise from a sample comprising a single outlier.

Furthermore, by setting $n = 2m$ and $i = m$ in equation (5.26), we obtain the following recurrence relation.

Relation 5.13: For $m = 1,2,\ldots,$

$$2\big[\mu_{m,m+1:2m} - \nu_{m,m+1:2m}\big]$$
$$= \sum_{r=0}^{m-1} \sum_{s=1}^{m-1} (-1)^{r+s-1} \binom{2m-1}{s} \binom{2m-1-s}{r} \Big\{\mu_{m-s,m-s+1:2m-r-s} - \nu_{m-s,m-s+1:2m-r-s}\Big\}$$
$$+ \sum_{r=1}^{m-1} (-1)^{r-1} \binom{2m-1}{r} \Big\{\mu_{m,m+1:2m-r} - \nu_{m,m+1:2m-r}\Big\}$$
$$+ \frac{1}{2} \binom{2m}{m} \Big\{\nu_{m:m}\big[\mu_{1:m} - \nu_{1:m}\big] + \nu_{1:m}\big[\mu_{m:m} - \nu_{m:m}\big]\Big\}. \tag{5.28}$$

In particular, for the case $m = 1$, upon using the result that $\nu_{1,2:2} = \nu_{1:1}^2$ we see that Relation 5.13 reduces to the well-known identity $\mu_{1,2:2} = \nu_{1:1}\mu_{1:1}$.

5.4. Relations for covariances

Making use of the results derived in Sections 2 and 3, we now obtain upper bounds for the number of single and product moments to be evaluated for calculating the means, variances and covariances of order statistics in a sample of size n by assuming these quantities to be available in samples of sizes $n-1$ and less. The following theorem, proved by Balakrishnan (1988a), thus generalizes the result given in Section 2.3 to the case when the order statistics arise from a sample containing a single outlier.

Theorem 5.14: In order to find the means, variances and covariances of order statistics in a sample of size n out of which $n-1$ variables are from an arbitrary continuous population with cdf $F(x)$ and one outlying variable from another continuous population with cdf $G(x)$, given these quantities in samples of sizes $n-1$ and less and also the quantities corresponding to the population with cdf $F(x)$, one has to evaluate at most two single moments and $(n-2)/2$ product moments for even values of n; and two single moments and $(n-1)/2$ product moments for odd values of n.

Proof. In view of Relation 5.4, it is sufficient to evaluate just two single moments $\mu_{n:n}$ and $\mu_{n:n}^{(2)}$ for calculating the means and variances of all order statistics in a sample of size n. Further, as mentioned earlier Relation 5.9 would enable us to calculate all the product moments $\mu_{i,j:n}$ $(1 \le i < j \le n)$ if we know n–1 suitably chosen moments, for example $\mu_{i,i+1:n}$ $(1 \le i \le n-1)$. When n is odd, we need to evaluate only (n–1)/2 of these moments as the remaining (n–1)/2 product moments could be obtained by applying Relation 5.11. Similarly, when n is even, say n = 2m, we need to evaluate only (n–2)/2 = m–1 of the immediate upper–diagonal product moments as Relation 5.13 gives the product moment $\mu_{m,m+1:2m}$ and the remaining m–1 product moments could be obtained by applying Relation 5.11.

Relation 5.15: For $1 \le k \le n-1$,

$$\sum_{j=2}^{n-k+1} \binom{n-j}{k-1} \mu_{1,j:n} + \sum_{j=2}^{k+1} \binom{n-j}{n-k-1} \mu_{1,j:n}$$
$$= \binom{n-1}{k} \nu_{1:k}\, \mu_{1:n-k} + \binom{n-1}{k-1} \nu_{1:n-k}\, \mu_{1:k}. \tag{5.29}$$

Proof. For $1 \le k \le n-1$, first consider

$$\sum_{j=2}^{n-k+1} \binom{n-j}{k-1} \mu_{1,j:n} = \sum_{j=2}^{n-k+1} \binom{n-j}{k-1} \iint_{W_1} xy\, h_{1,j:n}(x,y)\, dx\, dy, \tag{5.30}$$

where $h_{1,j:n}(x,y)$ is as given in equation (5.6). Now upon interchanging the summation and the integral signs and then using the binomial identity that

$$\sum_{r=0}^{m} \binom{m}{r} \{F(y)-F(x)\}^r \{1-F(y)\}^{m-r} = \{1-F(x)\}^m, \tag{5.31}$$

we obtain from (5.30) that

$$\sum_{j=2}^{n-k+1} \binom{n-j}{k-1} \mu_{1,j:n} = \iint_{W_1} xy\, H_{k,n}(x,y)\, dx\, dy, \tag{5.32}$$

where

$$\begin{aligned} H_{k,n}(x,y) &= \frac{(n-1)!}{(k-1)!(n-k-1)!} \{1-F(x)\}^{n-k-1}\{1-F(y)\}^{k-1} g(x)\, f(y) \\ &+ \frac{(n-1)!}{(k-1)!(n-k-1)!} \{1-F(x)\}^{n-k-1}\{1-F(y)\}^{k-1} f(x)\, g(y) \\ &+ \frac{(n-1)!}{(k-1)!(n-k-2)!} \{1-F(x)\}^{n-k-2}\{1-F(y)\}^{k-1} \{1-G(x)\}\, f(x)\, f(y) \\ &+ \frac{(n-1)!}{(k-2)!(n-k-1)!} \{1-F(x)\}^{n-k-1}\{1-F(y)\}^{k-2} \{1-G(y)\}\, f(x)\, f(y). \end{aligned} \tag{5.33}$$

Next, consider for $1 \le k \le n-1$

$$\sum_{j=2}^{k+1} \binom{n-j}{n-k-1} \mu_{i,j:n} = \sum_{j=2}^{k+1} \binom{n-j}{n-k-1} \iint_{W_2} xy\, h_{1,j:n}(y,x)\, dy\, dx, \tag{5.34}$$

from equation (5.15). Now upon interchanging the summation and the integral signs and then using the binomial identity that

$$\sum_{r=0}^{m} \binom{m}{r} \{F(x)-F(y)\}^r \{1-F(x)\}^{m-r} = \{1-F(y)\}^m,$$

we obtain from (5.34) that

$$\sum_{j=2}^{k+1} \binom{n-j}{n-k-1} \mu_{1,j:n} = \iint_{W_2} xy\, H_{k,n}(x,y)\, dx\, dy, \tag{5.35}$$

where $H_{k,n}(x,y)$ is as defined in (5.33). Finally, upon adding equations (5.32) and (5.35), noting that $W_1 \cup W_2 = \mathbb{R}^2$, and then simplifying the resulting expression by using equations (5.7) and (5.8), we obtain the relation in (5.29).

Note that Relation 5.15 involves the product moments $\mu_{1,j:n}$ $(2 \le j \le n)$ and first order single moments only, and that there are only [n/2] distinct equations since the relation for k is same as the relation for n–k. Thus, for even values of n, there are only n/2 equations in n–1 product moments and a knowledge of (n–2)/2 of these would enable us to calculate all of these product moments provided the first single moment in samples of sizes less than n are all known. Similarly, for odd values of n, we only need to know (n–1)/2 product moments. Note that these bounds are exactly same as the bounds given in Theorem 5.14 for the product moments to be evaluated for the calculation of all the product moments. This is quite expected, since the product moments $\mu_{1,j:n}$ $(2 \le j \le n)$, along with Relation 5.9 are also sufficient for the evaluation of all the product moments.

Relation 5.16: For $1 \le i \le n-1$,

$$\sum_{j=i+1}^{n} \mu_{i,j:n} + \sum_{r=1}^{i} \mu_{r,i+1:n} = (n-1)\, \nu_{1:1}\, \mu_{i:n-1} + \nu_{i:n-1}\, \mu_{1:1}. \tag{5.36}$$

Proof. For $1 \le i \le n-1$, consider

$$\sum_{j=i+1}^{n} \mu_{i,j:n} = \sum_{j=i+1}^{n} \iint_{W_1} xy\, h_{i,j:n}(x,y)\, dx\, dy, \tag{5.37}$$

where $h_{i,j:n}(x,y)$ is as given in (5.6). Upon interchanging the summation and the integral signs and then using the binomial identity in (5.31), we get from (5.37) that

$$\sum_{j=i+1}^{n} \mu_{i,j:n} = \iint_{W_1} xy\, H^*_{i,n}(x,y)\, dx\, dy, \tag{5.38}$$

where

$$\begin{aligned} H^*_{i,n}(x,y) &= \frac{(n-1)!}{(i-2)!(n-i-1)!} \{F(x)\}^{i-2} \{1-F(x)\}^{n-i-1} G(x)\, f(x)\, f(y) \\ &+ \frac{(n-1)!}{(i-1)!(n-i-1)!} \{F(x)\}^{i-1} \{1-F(x)\}^{n-i-1} g(x)\, f(y) \\ &+ \frac{(n-1)!}{(i-1)!(n-i-2)!} \{F(x)\}^{i-1} \{1-F(x)\}^{n-i-2} \{1-G(x)\}\, f(x)\, f(y) \\ &+ \frac{(n-1)!}{(i-1)!(n-i-1)!} \{F(x)\}^{i-1} \{1-F(x)\}^{n-i-1} f(x)\, g(y). \end{aligned} \tag{5.39}$$

Proceeding similarly, we also obtain

$$\begin{aligned} \sum_{r=1}^{i} \mu_{r,i+1:n} &= \sum_{r=1}^{i} \iint_{W_2} xy\, h_{r,i+1:n}(y,x)\, dy\, dx \\ &= \iint_{W_2} xy\, H^*_{i,n}(x,y)\, dx\, dy, \end{aligned} \tag{5.40}$$

where $H^*_{i,n}(x,y)$ is as defined above in (5.39). By adding equations (5.38) and (5.40), noting that $W_1 \cup W_2 = \mathbb{R}^2$, and then simplifying the resulting expression using equations (5.7) and (5.8), we derive the required relation.

Relation 5.16 has been made use of by Balakrishnan (1988a) in deriving some similar relations satisfied by the covariances of order statistics, viz., $\sigma_{i,j:n} = \mathrm{Cov}(Z_{i:n}, Z_{j:n}) = \mu_{i,j:n} - \mu_{i:n}\,\mu_{j:n}$ $(1 \le i < j - \le n)$. These generalize the results of Joshi and Balakrishnan (1982) to the case when the order statistics arise from a sample comprising a single outlier.

Relation 5.17: For $1 \le i \le n-1$,

$$\sum_{j=i+1}^{n} \sigma_{i,j:n} + \sum_{j=1}^{i} \sigma_{j,i+1:n}$$
$$= \left[i\,\nu_{1:1} - \sum_{j=1}^{i} \mu_{j:n}\right]\left[\mu_{i+1:n} - \mu_{i:n}\right] - \left[\mu_{1:1} - \nu_{1:1}\right]\left[\mu_{i:n} - \nu_{i:n-1}\right]. \tag{5.41}$$

Proof. Using the fact that $\mu_{i,j:n} = \sigma_{i,j:n} + \mu_{i:n}\,\mu_{j:n}$ in (5.36) we get

$$\sum_{j=i+1}^{n} \sigma_{i,j:n} + \sum_{j=1}^{i} \sigma_{j,i+1:n}$$
$$= (n-1)\nu_{1:1}\,\mu_{i:n-1} + \nu_{i:n-1}\,\mu_{1:1} - \mu_{i:n} \sum_{j=i+1}^{n} \mu_{j:n} - \mu_{i+1:n} \sum_{j=1}^{i} \mu_{j:n}. \tag{5.42}$$

With the identity

$$\sum_{j=i+1}^{n} \mu_{j:n} = (n-1)\,\nu_{1:1} + \mu_{1:1} - \sum_{j=1}^{i} \mu_{j:n}$$

obtained from Relation 5.1, we have the RHS of equation (5.42) as

$$(n-1)\,\nu_{1:1}\left[\mu_{i:n-1} - \mu_{i:n}\right] + \mu_{1:1}\left[\nu_{i:n-1} - \mu_{i:n}\right] - \left[\mu_{i+1:n} - \mu_{i:n}\right] \sum_{j=1}^{i} \mu_{j:n}.$$

Now making use of the result that

$$(n-1)\left[\mu_{i:n-1} - \mu_{i:n}\right] = i\left[\mu_{i+1:n} - \mu_{i:n}\right] - \left[\nu_{i:n-1} - \mu_{i:n}\right]$$

obtained from Relation 5.2, we derive the relation in (5.41).

Note that Relations 5.16 and 5.17 give extremely simple and useful results for checking the calculations of product moments and covariances of order statistics from a sample of size n comprising a single outlier. In particular, setting $i = 1$ and $i = n-1$ in equation (5.41), we get the identities

$$2\,\sigma_{1,2:n} + \sum_{j=3}^{n} \sigma_{1,j:n}$$
$$= \left[\nu_{1:1} - \mu_{1:n}\right]\left[\mu_{2:n} - \mu_{1:n}\right] - \left[\mu_{1:1} - \nu_{1:1}\right]\left[\mu_{1:n} - \nu_{1:n-1}\right]$$

and

$$2\,\sigma_{n-1,n:n} + \sum_{j=1}^{n-2} \sigma_{j,n:n} = \left[\mu_{n:n} - \mu_{1:1}\right]\left[\mu_{n:n} - \mu_{n-1:n}\right] - \left[\mu_{1:1} - \nu_{1:1}\right]\left[\mu_{n-1:n} - \nu_{n-1:n-1}\right]$$
$$= \left[\mu_{n:n} - \mu_{1:1}\right]\left[\mu_{n:n} - \nu_{n-1:n-1}\right] - \left[\mu_{n:n} - \nu_{1:1}\right]\left[\mu_{n-1:n} - \nu_{n-1:n-1}\right].$$

The last equation has been obtained from the previous equation simply by rearranging the terms on the right–hand side.

5.5. Results for symmetric outlier model

Let us assume that the density functions f(x) and g(x) are both symmetric about zero. It is then easy to see from equations (5.4) and (5.6) that

$$h_{n-i+1:n}(x) = h_{i:n}(-x), \qquad 1 \le i \le n,$$

and

$$h_{n-j+1,n-i+1:n}(y,x) = h_{i,j:n}(-x,-y), \qquad 1 \le i < j \le n.$$

As a result, for a symmetric outlier model (the case when both f(x) and g(x) are symmetric about zero) we have the relations

$$\mu^{(k)}_{n-i+1:n} = (-1)^k\, \mu^{(k)}_{i:n}, \qquad 1 \le i \le n, \quad k \ge 1, \tag{5.43}$$

$$\mu_{n-j+1,n-i+1:n} = \mu_{i,j:n}, \qquad 1 \le i < j \le n, \tag{5.44}$$

and

$$\begin{aligned}\sigma_{n-j+1,n-i+1:n} &= \mu_{n-j+1,n-i+1:n} - \mu_{n-j+1:n}\,\mu_{n-i+1:n} = \mu_{i,j:n} - \mu_{i:n}\,\mu_{j:n} \\ &= \sigma_{i,j:n}, \qquad 1 \le i \le j \le n.\end{aligned} \tag{5.45}$$

Equations (5.43) – (5.45) could be used to simplify many of the results presented in Sections 2 through 4. These relations then would help us reduce the bounds given in Theorem 5.14 for the case when the order statistics arise from a symmetric outlier model.

Relation 5.18: For n even, say n = 2m,

$$\mu^{(k)}_{m:2m-1} = 0 \quad \text{for odd } k \tag{5.46}$$

$$= \frac{1}{2m-1}\left\{2m\,\mu^{(k)}_{m:2m} - \nu^{(k)}_{m:2m-1}\right\} \quad \text{for even } k. \tag{5.47}$$

Proof. Equation (5.46) is obtained from (5.43) simply by setting n = 2m–1 and i = m. Equation (5.47), on the other hand, follows directly from Relation 5.3 upon using the result that $\mu^{(k)}_{m+1:2m} = \mu^{(k)}_{m:2m}$ for even values of k.

Next, upon using (5.43) and (5.44) in Relation 5.10 we also derive the following result.

Relation 5.19: For $1 \le i < j \le n$,

$$\begin{aligned}(1+(-1)^n)\left[\mu_{i,j:n} - \nu_{i,j:n}\right] &= \frac{1}{n}\sum_{r=1}^{j-i} (-1)^{j-i-r-1}\binom{n}{j-r}\binom{j-r-1}{i-1}\Big\{(j-r)\,\nu_{r:n-j+r}\left[\mu_{1:j-r} - \nu_{1:j-r}\right] \\ &\quad + (n-j+r)\,\nu_{1:j-r}\left[\mu_{r:n-j+r} - \nu_{r:n-j+r}\right]\Big\} \\ &\quad + \sum_{r=1}^{i-1} (-1)^{n-r-1}\binom{n-1}{r}\left\{\mu_{n-j+1,n-i+1:n-r} - \nu_{n-j+1,n-i+1:n-r}\right\} \\ &\quad + \sum_{r=0}^{i-1}\sum_{s=1}^{n-j} (-1)^{n-r-s-1}\binom{n-1}{s}\binom{n-1-s}{r}\Big\{\mu_{n-j-s+1,n-i-s+1:n-r-s} \\ &\qquad - \nu_{n-j-s+1,n-i-s+1:n-r-s}\Big\}.\end{aligned} \tag{5.48}$$

Note that the right–hand side of the above realtion involves only the expected values and product moments in samples of sizes n–1 and less. In particular, for even values of n we get from (5.48) that

$$2\,\mu_{i,j:n} = 2\,\mu_{n-j+1,n-i+1:n}$$
$$= 2\,\nu_{i,j:n} + \sum_{r=1}^{i-1} (-1)^{r-1} \binom{n-1}{r} \left\{\mu_{n-j+1,n-i+1:n-r} - \nu_{n-j+1,n-i+1:n-r}\right\}$$
$$+ \frac{1}{n} \sum_{r=1}^{j-i} (-1)^{j-i-r-1} \binom{n}{j-r} \binom{j-r-1}{i-1} \Big\{(j-r)\,\nu_{r:n-j+r} \left[\mu_{1:j-r} - \nu_{1:j-r}\right]$$
$$+ (n-j+r)\,\nu_{1:j-r} \left[\mu_{r:n-j+r} - \nu_{r:n-j+r}\right]\Big\}$$
$$+ \sum_{r=0}^{i-1} \sum_{s=1}^{n-j} (-1)^{r+s-1} \binom{n-1}{s} \binom{n-1-s}{r} \left\{\mu_{n-j-s+1,n-i-s+1:n-r-s} - \nu_{n-j-s+1,n-i-s+1:n-r-s}\right\}. \qquad (5.49)$$

This result, established by Balakrishnan (1988a) generalizes the relation in (2.43) to the case when the order statistics arise from a symmetric outlier model. Also from equation (5.49) we may note that we do not have to evaluate any product moments in a sample of size n whenever n is even. In addition, by setting i = n–1 and j = n in equation (5.49) and using the fact that $\mu_{1:1} = \nu_{1:1} = 0$, we get

$$2\,\mu_{1,2:n} = 2\,\nu_{1,2:n} + \sum_{r=1}^{n-2} (-1)^{r-1} \binom{n-1}{r} \left\{\mu_{1,2:n-r} - \nu_{1,2:n-r}\right\}$$

which, upon using the result in (2.44) yields the relation

$$2\,\mu_{1,2:n} = \sum_{r=1}^{n-2} (-1)^{r-1} \left\{\binom{n-1}{r} \mu_{1,2:n-r} + \binom{n-1}{r-1} \nu_{1,2:n-r}\right\} \qquad (5.50)$$

for even values of n.

In the following theorem, which is an analogue of Theorem 5.14 for the symmetric outlier model case, we essentially make use of equations (5.46) – (5.49) in order to determine the upper bounds for the number of single and product moments to be evaluated in a sample of size n.

Theorem 5.20: In order to determine the means, variances and covariances of order statistics in a sample of size n out of which n–1 variables are from an arbitrary continuous symmetric population with cdf F(x) and one outlying variable from another continuous symmetric population with cdf G(x), given these quantities in samples of sizes n–1 and less and also the quantities corresponding to the population with cdf F(x), one has to evaluate at most one single moment if n is even; and one single moment and (n–1)/2 product moments if n is odd.

Proof. In order to compute the first single moments $\mu_{i:n}$ $(1 \le i \le n)$, because of Relation 5.4 we have to evaluate at most one single moment if n is even and no single moment if n is odd as in this case we have $\mu_{\frac{n+1}{2}:n} = 0$ from equation (5.46). Next, in order to compute the second single moments $\mu^{(2)}_{i:n}$ $(1 \le i \le n)$, we have to evaluate at most one single moment if n is odd and no single moment if n is even as in this case we have

$$\mu^{(2)}_{\frac{n}{2}:n} = \frac{1}{n}\left\{(n-1)\,\mu^{(2)}_{\frac{n}{2}:n-1} + \nu^{(2)}_{\frac{n}{2}:n-1}\right\}$$

from equation (5.47). Finally, in order to obtain all the product moments $\mu_{i,j:n}$ $(1 \le i < j \le n)$, we note from Theorem 5.14 that we have to evaluate at most (n–1)/2 product moments if n is odd; however, for even values of n we do not have to evaluate any product moment as in this case we could compute all the product moments by using equation (5.49). Hence, the theorem.

5.6. Results for two related outlier models

In the previous section we have established some relations for both single and product moments of order statistics from a symmetric outlier model. In this section we once again consider the moments of order statistics from a symmetric outlier model and express them in terms of the moments of order statistics in samples drawn from the population with pdf $f^*(x)$ (obtained by folding the pdf f(x) at zero) and the moments of order statistics in samples drawn from the population with pdf $f^*(x)$ comprising a single outlier with pdf $g^*(x)$ (obtained by folding the pdf g(x) at zero). These results have been proved recently by Balakrishnan (1988b) and they generalize the relations presented in Section 2.6 to the case when the order statistics arise from a symmetric outlier model. These results have also been successfully applied by Balakrishnan and Ambagaspitiya (1988) in order to evaluate the means, variances and covariances of order statistics from a single scale–outlier double exponential model. They have then used these quantities in order to examine the variances of various location estimators expressible as linear functions of the order statistics. Similar work for the normal case has been carried out by David and Shu (1978).

In this regard, let us denote for $x > 0$

$$F^*(x) = 2\,F(x) - 1, \qquad f^*(x) = 2\,f(x), \tag{5.51}$$

and

$$G^*(x) = 2\,G(x) - 1, \qquad g^*(x) = 2\,g(x). \tag{5.52}$$

That is, the density functions $f^*(x)$ and $g^*(x)$ are obtained by folding the density functions f(x) and g(x) at the point zero, respectively. Let us now denote the single and the product moments of order statistics in a random sample of size n drawn from a population with pdf $f^*(x)$ and cdf $F^*(x)$ defined in (5.51) by $\nu_{i:n}^{*(k)}$ $(1 \le i \le n,\ k \ge 1)$ and $\nu_{i,j:n}^{*}$ $(1 \le i < j \le n)$. Furthermore, let $\mu_{i:n}^{*(k)}$ $(1 \le i \le n,\ k \ge 1)$ and $\mu_{i,j:n}^{*}$ $(1 \le i < j \le n)$ denote the single and the product moments of order statistics obtained from a sample of n independent random variables out of which n–1 have pdf $f^*(x)$ and cdf $F^*(x)$ defined in (5.51) and one variable has pdf $g^*(x)$ and cdf $G^*(x)$ defined in (5.52). Then Balakrishnan (1988b) has essentially derived some simple formulae expressing the moments $\mu_{i:n}^{(k)}$, $\mu_{i,j:n}$ in terms of $\mu_{i:n}^{*(k)}$, $\nu_{i:n}^{*(k)}$, $\mu_{i,j:n}^{*}$ and $\nu_{i,j:n}^{*}$.

<u>Relation 5.21</u>: For $1 \le i \le n$ and $k = 1,2,\ldots,$

$$2^n\,\mu_{i:n}^{(k)} = \sum_{r=1}^{i-1} \binom{n-1}{r-1} \nu_{i-r:n-r}^{*(k)} + (-1)^k \sum_{r=i}^{n-1} \binom{n-1}{r} \nu_{r-i+1:r}^{*(k)}$$

$$+\sum_{r=0}^{i-1}\binom{n-1}{r}\mu^{*(k)}_{i-r:n-r}+(-1)^k\sum_{r=i}^{n}\binom{n-1}{r-1}\mu^{*(k)}_{r-i+1:r}. \tag{5.53}$$

Proof. First, express the single moment $\mu^{(k)}_{i:n}$ as the sum of three integrals using equations (5.4) and (5.7) and split the integrals each into two parts, one on the range $(-\infty,0)$ and the other on the range $(0,\infty)$. In the integrals on the range $(-\infty,0)$, make a substitution such that the range becomes $(0,\infty)$, use the facts $F(-x) = 1-F(x)$ and $G(-x) = 1-G(x)$ along with equations (5.51) and (5.52), and express the integrands in terms of F^*, G^*, f^* and g^*. Now expand $\{1+F^*(x)\}^m$ in powers of $F^*(x)$, integrate termwise and the relation in (5.53) readily follows.

A similar relation for the product moments is established in the following result.

Relation 5.22: For $1 \le i < j \le n$,

$$\begin{aligned}2^n\mu_{i,j:n} &= \sum_{r=1}^{i-1}\binom{n-1}{r-1}\nu^*_{i-r,j-r:n-r}+\sum_{r=j}^{n-1}\binom{n-1}{r}\nu^*_{r+1-j,r+1-i:r}\\ &+\sum_{r=0}^{i-1}\binom{n-1}{r}\mu^*_{i-r,j-r:n-r}+\sum_{r=j}^{n}\binom{n-1}{r-1}\mu^*_{r+1-j,r+1-i:r}\\ &-\sum_{r=i}^{j-1}\binom{n-1}{r-1}\nu^*_{j-r:n-r}\,\mu^*_{r+1-i:r}-\sum_{r=i}^{j-1}\binom{n-1}{r}\nu^*_{r+1-i:r}\,\mu^*_{j-r:n-r}.\end{aligned} \tag{5.54}$$

Proof. First of all, express the product moment $\mu_{i,j:n}$ as the sum of five integrals using equations (5.6) and (5.14). Next, by noting that

$W_1 = \{(x,y):\ -\infty < x < y < \infty\} = R_1 \cup R_2 \cup R_3,$

where

$R_1 = \{(x,y):\ 0 < x < y < \infty\}$, $R_2 = \{(x,y):\ -\infty < x < 0, 0 < y < \infty\}$

and

$R_3 = \{(x,y):\ -\infty < x < y < 0\},$

split each of the five integrals into three parts. In the integrals on the range R_1, express F and G in terms of F^* and G^* and f and g in terms of f^* and g^*, respectively, expand $(1 + F^*)^m$ binomially in powers of F^* and integrate termwise. In the integrals on the range R_2, make a substitution $x = -z$, use the results that $F(-x) = 1-F(x)$ and $G(-x) = 1-G(x)$, express the integrand in terms of F^*, G^*, f^* and g^*, expand $\{F^*(y) + F^*(x)\}^m$ binomially in powers of $F^*(y)$ and $F^*(x)$ and then integrate termwise. Finally in the integrals on the range R_3, put $x = -u$ and $y = -v$, use the results that $F(-z) = 1-F(z)$ and $G(-z) = 1-G(z)$, express the integrand in terms of F^*, G^*, f^* and g^*, expand $(1+F^*)^m$ binomially in powers of F^* and integrate termwise. Combining all these results and simplifying the resulting expression, we derive the relation in (5.54).

Remark 5.23: If the moments $\nu^{*(k)}_{i:m}$, $\mu^{*(k)}_{i:m}$, $\nu^*_{i,j:m}$ and $\mu^*_{i,j:m}$ are all available for sample sizes up to n, then all the single moments $\mu^{(k)}_{i:n}$ $(1 \le i \le n)$ and the product moments $\mu_{i,j:n}$ $(1 \le i < j \le n)$ of order statistics in a sample of size n from a symmetric outlier model (with a single outlier present) could be computed by using Relations 5.21 and 5.22. Thus, for example, Balakrishnan and Ambagaspitiya (1988) have applied Relations 5.21 and 5.22 in computing the means, variances and covariances of order statistics from a single scale–outlier double exponential model by making use of the single and the product moments of order statistics from a standard exponential distribution and also the single and the product moments of order statistics from a single scale–outlier exponential model.

Remark 5.24: In particular, if we set $G(x) \equiv F(x)$ (that is, when the variable Y in the sample is not an outlier), then Relations 5.21 and 5.22 simply reduce to

$$2^n \nu_{i:n}^{(k)} = \sum_{r=0}^{i-1} \binom{n}{r} \nu_{i-r:n-r}^{*(k)} + (-1)^k \sum_{r=i}^{n} \binom{n}{r} \nu_{r+1-i:r}^{*(k)}$$

and

$$2^n \nu_{i,j:n} = \sum_{r=0}^{i-1} \binom{n}{r} \nu_{i-r,j-r:n-r}^{*} + \sum_{r=j}^{n} \binom{n}{r} \nu_{r+1-j,r+1-i:r}^{*} - \sum_{r=i}^{j-1} \binom{n}{r} \nu_{r+1-i:r}^{*} \nu_{j-r:n-r}^{*},$$

respectively. Note that these are precisely the same as the results presented in Section 2.6.

Remark 5.25: In using Relations 5.21 and 5.22, an error coud essentially arise from two sources, (i) due to approximations in the coefficients, and (ii) due to approximations in the pivotal quantities. Since the coefficients occurring in both the relations are simple binomial coefficients which are integral and could be evaluated exactly at least for small values of n, we may assume that the error involved due to the approximations in the coefficients to be zero. Now if ϵ and ϵ' are the errors involved in approximating each of the pivotal values $\nu^{*(k)}$ and $\mu^{*(k)}$, respectively, then the maximum cumulative rounding error in evaluating $\mu_{i:n}^{(k)}$ $(1 \le i \le n)$ by means of Relation 5.21 is given by

$$2^{-n} \left[\epsilon \sum_{r=1}^{i-1} \binom{n-1}{r-1} + \epsilon \sum_{r=i}^{n-1} \binom{n-1}{r} + \epsilon' \sum_{r=0}^{i-1} \binom{n-1}{r} + \epsilon' \sum_{r=i}^{n} \binom{n-1}{r-1} \right]$$

$$< 2^{-n} \epsilon^* \left[\sum_{r=0}^{n-1} \binom{n-1}{r} + \sum_{r=1}^{n} \binom{n-1}{r-1} \right] < 2^{-(n-1)} \epsilon^* \sum_{r=0}^{n-1} \binom{n-1}{r} = \epsilon^*,$$

where $\epsilon^* = \max(\epsilon,\epsilon')$. That is, the maximum error involved in numerically computing the single moments $\mu_{i:n}^{(k)}$ $(1 \le i \le n,\ k \ge 1)$ using the moments $\nu^{*(k)}$ and $\mu^{*(k)}$ is at most $\epsilon^* = \max(\epsilon,\epsilon')$, where ϵ and ϵ' are respectively the maximum errors involved in approximating each of the pivotal quantities $\nu^{*(k)}$ and $\mu^{*(k)}$. Hence, if the moments $\nu_{i:m}^{*(k)}$, $\mu_{i:m}^{*(k)}$, $\nu_{i,j:m}^{*}$ and $\mu_{i,j:m}^{*}$ are computed sufficiently accurately, then the single moments $\mu_{i:n}^{(k)}$ $(1 < i \le n,\ k \ge 1)$ and the product moments $\mu_{i,j:n}$ $(1 \le i < j \le n)$ could all be computed by using Relations 5.21 and 5.22 without accumulating serious rounding errors, at least for small sample sizes.

5.7. Functional behaviour of order statistics

Let us first consider the special case of a location–outlier model, i.e., $G(x) = F(x-\lambda)$ for all x. We could now write $Y = X_n+\lambda$, where X_n is a random variable with pdf f(x) and F(x) and independent of the remaining n−1 variables $X_1, X_2,\ldots,X_{n-1}$. For convenience let us denote the i'th order statistic in this case by $Z_{i:n}(\lambda)$, its pdf by $h_{i:n}(x;\lambda)$ and the cdf by $H_{i:n}(x;\lambda)$. Then it could be easily seen from equation (5.3) that $H_{i:n}(x;\lambda)$ is a decreasing function of λ. The behaviour of $Z_{i:n}(\lambda)$ as a function of λ has been studied in detail by David and Shu (1978) (also see Hampel, 1974). Denoting the observed values of the random variables X, Y and Z by x, y and z, and inserting $y = x_n+\lambda$ into the ordered sample of size n–1, viz., $x_{1:n-1} \le x_{2:n-1} \le \dots \le x_{n-1:n-1}$, we then have for fixed values of $x_1, x_2, \dots, x_n$ that

$$z_{1:n}(\lambda) = \begin{cases} x_n+\lambda & \text{if } x_n+\lambda \le x_{1:n-1} \\ x_{1:n-1} & \text{if } x_n+\lambda > x_{1:n-1} \end{cases} \tag{5.55}$$

and for i = 2,3,...,n–1

$$z_{i:n}(\lambda) = \begin{cases} x_{i-1:n-1} & \text{if } x_n+\lambda \le x_{i-1:n-1} \\ x_n+\lambda & \text{if } x_{i-1:n-1} < x_n+\lambda \le x_{i:n-1} \\ x_{i:n-1} & \text{if } x_n+\lambda > x_{i:n-1} \end{cases} \tag{5.56}$$

and

$$z_{n:n}(\lambda) = \begin{cases} x_{n-1:n-1} & \text{if } x_n+\lambda \le x_{n-1:n-1} \\ x_n+\lambda & \text{if } x_n+\lambda > x_{n-1:n-1}. \end{cases} \tag{5.57}$$

From equations (5.55) – (5.57), we see that $z_{i:n}(\lambda)$ is a nondecreasing function of λ; also

$$\lim_{\lambda\to-\infty} z_{1:n}(\lambda) = z_{1:n}(-\infty) = -\infty,$$

$$\lim_{\lambda\to-\infty} z_{i:n}(\lambda) = z_{i:n}(-\infty) = x_{i-1:n-1}, \qquad 2 \le i \le n,$$

$$\lim_{\lambda\to\infty} z_{i:n}(\lambda) = z_{i:n}(\infty) = x_{i:n-1}, \qquad 1 \le i \le n-1,$$

and

$$\lim_{\lambda\to\infty} z_{n:n}(\lambda) = z_{n:n}(\infty) = \infty. \tag{5.58}$$

For a finite value of λ we see from equations (5.55) – (5.57) that $E(Z_{i:n}(\lambda)) = \mu_{i:n}(\lambda)$, $1 \le i \le n$, exists if E(X) exists. By using the monotone convergence theorem (for example, see, Loeve, 1977) we also get for $2 \le i \le n$

$$\lim_{\lambda\to-\infty} E(Z_{i:n}(\lambda)) = E\left[\lim_{\lambda\to-\infty} Z_{i:n}(\lambda)\right]$$

implying that

$$\mu_{i:n}(-\infty) = E(X_{i-1:n-1}) = \nu_{i-1:n-1}; \tag{5.59}$$

similarly for $1 \le i \le n-1$

$$\lim_{\lambda\to\infty} E(Z_{i:n}(\lambda)) = E\left[\lim_{\lambda\to\infty} Z_{i:n}(\lambda)\right]$$

implying that

$$\mu_{i:n}(\infty) = E(X_{i:n-1}) = \nu_{i:n-1}. \tag{5.60}$$

Also

$$\mu_{1:n}(-\infty) = -\infty \text{ and } \mu_{1:n}(\infty) = \infty. \tag{5.61}$$

Furthermore, upon noting that for fixed x and y

$$f(x-\lambda) \to 0,\ F(x-\lambda) \to 0,\ F(y-\lambda) - F(x-\lambda) \to 0$$

as $\lambda \to \infty$, we have from equations (5.4) and (5.6) that

$$\lim_{\lambda\to\infty} h_{i:n}(x;\lambda) = h_{i:n}(x;\infty) = f_{i:n-1}(x), \qquad 1 \le i \le n-1, \tag{5.62}$$

and

$$\lim_{\lambda\to\infty} h_{i,j:n}(x,y;\lambda) = h_{i,j:n}(x,y;\infty) = f_{i,j:n-1}(x,y), \qquad 1 \le i < j \le n-1. \tag{5.63}$$

Using a similar argument we also have

$$\lim_{\lambda\to-\infty} h_{i:n}(x;\lambda) = h_{i:n}(x;-\infty) = f_{i-1:n-1}(x), \quad 2 \le i \le n, \tag{5.64}$$

and

$$\lim_{\lambda\to-\infty} h_{i,j:n}(x,y;\lambda) = h_{i,j:n}(x,y;-\infty) = f_{i-1,j-1:n-1}(x,y), \quad 2 \le i < j \le n. \tag{5.65}$$

Now upon using the Lebesgue dominated convergence theorem (see Loeve, 1977) we obtain

$$\lim_{\lambda\to\infty} \sigma_{i,j:n}(\lambda) = \sigma_{i,j:n}(\infty) = \mathrm{Cov}(X_{i:n-1},X_{j:n-1}), \quad 1 \le i \le j \le n-1, \tag{5.66}$$

and

$$\lim_{\lambda\to-\infty} \sigma_{i,j:n}(\lambda) = \sigma_{i,j:n}(-\infty) = \mathrm{Cov}(X_{i-1:n-1},X_{j-1:n-1}), \quad 2 \le i \le j \le n. \tag{5.67}$$

Next, let us consider a scale–outlier model, i.e., $G(x) = F(x/\tau)$ for all x, where $\tau > 0$. In this case we could write $Y = \tau X_n$, where X_n is a random variable with pdf f(x) and cdf F(x) and independent of the n–1 variates $X_1,X_2,\ldots,X_{n-1}$. For convenience let us denote the i'th order statistic in this case by $Z^*_{i:n}(\tau)$, its pdf by $h^*_{i:n}(x;\tau)$ and the cdf by $H^*_{i:n}(x;\tau)$. From equation (5.3) it could then be seen that $H^*_{i:n}(x;\tau)$ is a decreasing function of τ for fixed positive values of x and an increasing function of τ for fixed negative values of x. The functional behaviour of $Z^*_{i:n}(\tau)$ as a function of τ has been studied by David and Shu (1978). As before, denoting the realizations of the variables X, Y and Z by x, y and z, and inserting $y = \tau x_n$ into the ordered sample $x_{1:n-1} \le x_{2:n-1} \le \ldots \le x_{n-1:n-1}$, we have for fixed values of $x_1,x_2,\ldots,x_n$ that

$$\begin{aligned} z^*_{1:n}(\tau) &= \tau x_n && \text{if } \tau x_n \le x_{1:n-1} \\ &= x_{1:n-1} && \text{if } \tau x_n > x_{1:n-1} \end{aligned} \tag{5.68}$$

and for $2 \le i \le n-1$

$$\begin{aligned} z^*_{i:n}(\tau) &= x_{i-1:n-1} && \text{if } \tau x_n \le x_{i-1:n-1} \\ &= \tau x_n && \text{if } x_{i-1:n-1} < \tau x_n \le x_{i:n-1} \\ &= x_{i:n-1} && \text{if } \tau x_n > x_{i:n-1} \end{aligned} \tag{5.69}$$

and

$$\begin{aligned} z^*_{n:n}(\tau) &= x_{n-1:n-1} && \text{if } \tau x_n \le x_{n-1:n-1} \\ &= \tau x_n && \text{if } \tau x_n > x_{n-1:n-1}. \end{aligned} \tag{5.70}$$

From equations (5.68) – (5.70), we see that $z^*_{i:n}(\tau)$ is nondecreasing in τ if $x_n > 0$ and is nonincreasing in τ if $x_n < 0$. Moreover, for $2 \le i \le n-1$,

$$\begin{aligned} \lim_{\lambda\to\infty} z^*_{i:n}(\tau) = z^*_{i:n}(\infty) &= x_{i:n-1} && \text{if } x_n > 0 \\ &= x_{i-1:n-1} && \text{if } x_n < 0. \end{aligned} \tag{5.71}$$

Now if X is symmetrically distributed about zero, we have from equation (5.71) that for $2 \le i \le n-1$

$$\begin{aligned} \lim_{\lambda\to\infty} E(Z^*_{i:n}(\tau)) &= E\left[\lim_{\tau\to\infty} Z^*_{i:n}(\tau)\right] = E\left[Z^*_{i:n}(\infty)\right] \\ &= \Pr\left[X_n > 0\right] E\left[X_{i:n-1}\right] + \Pr\left[X_n < 0\right] E\left[X_{i-1:n-1}\right] \end{aligned}$$

which implies

$$\mu^*_{i:n}(\infty) = \frac{1}{2}\left[\nu_{i-1:n-1} + \nu_{i:n-1}\right]. \tag{5.72}$$

In addition, we also have

$$\lim_{\tau\to\infty} E(Z^*_{1:n}(\tau)) = \mu^*_{1:n}(\infty) = -\infty$$

and

$$\lim_{\tau\to\infty} E(Z^*_{n:n}(\tau)) = \mu^*_{n:n}(\infty) = \infty.$$

We shall make use of all these results in the next section in order to examine the bias and the mean square error of various estimators of the location parameter that are expressible as linear functions of the order statistics when an unidentified single outlying observation is present in a sample of size n.

5.8. Applications in robustness studies

The robust estimation of the location and scale parameters of symmetric populations, in particular, normal distribution, has been of considerable interest in the recent past. For various symmetric populations, Crow and Siddiqui (1967) have studied the efficiency of some estimators of the location parameter such as the median, Winsorized means and trimmed means. Based on Monte Carlo methods, Andrews et al. (1972) have carried out a similar study for a much larger class of robust estimators of the location parameter that also includes many adaptive estimators which essentially adapt themselves to some special features of a given sample. One could also refer to Tiku et al. (1986) for a detailed comparative study of various robust estimators of the location and scale parameters.

Here we shall restrict our attention to the following estimators of the location parameter that are based on a sample of size n.

(a) Sample mean:

$$\bar{X}_n = \frac{1}{n}\sum_{i=1}^{n} Z_{i:n};$$

(b) Trimmed means:

$$T_n(r) = \frac{1}{n-2r}\sum_{i=r+1}^{n-r} Z_{i:n};$$

(c) Winsorized means:

$$W_n(r) = \frac{1}{n}\left[\sum_{i=r+2}^{n-r-1} Z_{i:n} + (r+1)\left[Z_{r+1:n} + Z_{n-r:n}\right]\right];$$

(d) Modified maximum likelihood estimators:

$$M_n(r) = \frac{1}{m}\left[\sum_{i=r+2}^{n-r-1} Z_{i:n} + (1+r\beta)\left[Z_{r+1:n} + Z_{n-r:n}\right]\right],$$

where $m = n-2r+2r\beta$; Tiku (1967, 1980) has given the expression for β while Tiku et al. (1986) have tabulated the values of β for various choices of n and r;

(e) Linearly weighted means:

$$L_n(r) = \frac{1}{2(\frac{n}{2}-r)^2}\sum_{i=1}^{\frac{n}{2}-r} (2i-1)\left[Z_{r+i:n} + Z_{n-r-i+1:n}\right]$$

for even values of n;

(f) Gastwirth mean:

$$G_n = 0.3\,(Z_{[\frac{n}{3}]+1:n} + Z_{n-[\frac{n}{3}]:n}) + 0.2\,(Z_{\frac{n}{2}:n} + Z_{\frac{n}{2}+1:n})$$

for even values of n, where $[\frac{n}{3}]$ denotes the integral part of $\frac{n}{3}$.

For the location–outlier model, the estimators considered above may all be written as

$$M(\lambda) = \sum_{i=1}^{n} a_i\, Z_{i:n}(\lambda).$$

From equations (5.55) – (5.57) it could be immediately seen that $\sum_{i=1}^{n} a_i\, z_{i:n}(\lambda)$ is a nondecreasing continuous function of λ and, in addition, $E(M(\lambda)) = \sum_{i=1}^{n} a_i\, \mu_{i:n}(\lambda)$ is an increasing function of λ, with $E(M(\infty)) = \infty$ except when $a_n = 0$, and $E(M(-\infty)) = -\infty$ except when $a_1 = 0$. From equations (5.58) and (5.59), we get when $a_1 = 0$ that

$$E(M(-\infty)) = \sum_{i=2}^{n} a_i\, v_{i-1:n-1},$$

and similarly when $a_n = 0$ we get from equations (5.58) and (5.60) that

$$E(M(\infty)) = \sum_{i=1}^{n-1} a_i\, v_{i:n-1}.$$

Making use of the tables of expected values of normal order statistics under the location–outlier model prepared by David et al. (1977), bias of various estimators of the mean μ of a normal N(μ,1) population, based on a sample of size n = 10 with one observation being from a normal N(μ+λ,1) distribution, have been computed for some specific values of λ. These are presented in Table 5.1 given below.

Table 5.1

Bias of various estimators of μ for n = 10 when a single observation is from N(μ+λ,1) and the others from N(μ,1)

	λ							
Estimator	0.0	0.5	1.0	1.5	2.0	3.0	4.0	∞
$\bar{X}_{10}$0.0	0.0	0.05000	0.10000	0.15000	0.20000	0.30000	0.40000	∞
$T_{10}(1)$	0.0	0.04912	0.09325	0.12870	0.15400	0.17871	0.18470	0.18563
$T_{10}(2)$	0.0	0.04869	0.09023	0.12041	0.13904	0.15311	0.15521	0.15538
Med_{10}	0.0	0.04832	0.08768	0.11381	0.12795	0.13642	0.13723	0.13726
$W_{10}(1)$	0.0	0.04938	0.09506	0.13368	0.16298	0.19407	0.20239	0.20377
$W_{10}(2)$	0.0	0.04889	0.09156	0.12389	0.14497	0.16217	0.16504	0.16530
$M_{10}(1)$	0.0	0.04934	0.09484	0.13311	0.16194	0.19229	0.20037	0.20169
$M_{10}(2)$	0.0	0.04886	0.09137	0.12342	0.14418	0.16091	0.16369	0.16394
$L_{10}(1)$	0.0	0.04869	0.09024	0.12056	0.13954	0.15459	0.15727	0.15758
$L_{10}(2)$	0.0	0.04850	0.08892	0.11700	0.13328	0.14436	0.14576	0.14585
G_{10}	0.0	0.04847	0.08873	0.11649	0.13237	0.14285	0.14407	0.14414

By looking at the values of the bias of various estimators given in Table 5.1, we get the ordering

$$\bar{X}_{10} \prec W_{10}(1) \prec M_{10}(1) \prec T_{10}(1) \prec W_{10}(2) \prec M_{10}(2) \prec L_{10}(1) \prec T_{10}(2) \prec L_{10}(2) \prec G_{10} \prec \text{Med}_{10}, \qquad (5.73)$$

where $\prec$ denotes "inferior to". It should be noted that the trimmed means have a smaller bias than the corresponding modified maximum likelihood estimators which, in turn, have a smaller bias than the corresponding Winsorized means. Also, as rightly pointed out by David and Shu (1978), the median is more biased than what we may naively have thought it to be. In addition, the estimators based on the 6 central order statistics are seen to be less subject to bias than those based on the 8 central order statistics. This is not surprising as we would expect an estimator omitting the 4 extreme observations (two smallest and two largest) to exclude the single outlier present in the sample with a larger probability as compared to an estimator omitting only the 2 extreme observations (one smallest and one largest). For the case when $n = 10$ and $\lambda = 2$, for example, we have the following values of $\Pr\{\text{rank}(Y)=i\} = \Pi_i$ from an extensive set of tables prepared by Milton (1970):

i:	1	2	3	4	5	6	7	8	9	10
Π_i:	0.001	0.003	0.005	0.009	0.014	0.023	0.040	0.072	0.159	0.674

From these values we see that an estimator omitting the four extreme observations (two smallest and two largest) excludes the outlier with probability 0.837 while an estimator omitting the two extreme observations (one smallest and one largest) excludes the outlier with probability 0.675 only.

Considering now the expression of $E(M^2(\lambda))$ and then integrating by parts, we obtain

$$E(M^2(\lambda)) = 2 \int_0^\infty x \Pr\{|M(\lambda)| > x\}\, dx. \qquad (5.74)$$

Assuming f to be a standardized symmetric density function and taking accordingly $a_i = a_{n-i+1}$ for $i = 1,2,\dots,[\frac{n+1}{2}]$, we have $M(0)$ to be symmetrically distributed about 0 and also $M(-\lambda)$ to be distributed exactly as $-M(\lambda)$. Then

$$\Pr\{|M(\lambda)| > x\} = \Pr\{M(\lambda) > x\} + \Pr\{M(-\lambda) > x\}$$

may be expected to be an increasing function of $|\lambda|$ and, consequently, we have $E(M^2(\lambda))$ to be an increasing function of λ from equation (5.74). Furthermore, for $a_1 = a_n = 0$, we obtain

$$\begin{aligned}
E(M^2(\pm\infty)) &= \lim_{\lambda\to\infty} E(M(\lambda))^2 \\
&= E\left[\lim_{\lambda\to\infty} \sum_{i=2}^{n-1} a_i Z_{i:n}(\lambda)\right]^2 \\
&= E\left[\sum_{i=2}^{n-1} a_i \lim_{\lambda\to\infty} Z_{i:n}(\lambda)\right]^2 \\
&= E\left[\sum_{i=2}^{n-1} a_i X_{i:n-1}\right]^2 \\
&= \left[E(M(\infty))\right]^2 + \sum_{i=2}^{n-1}\sum_{j=2}^{n-1} a_i a_j \operatorname{Cov}\left[X_{i:n-1}, X_{j:n-1}\right].
\end{aligned}$$

In Table 5.2 we have computed the mean square error of the various estimators considered earlier for whom the bias are given in Table 5.1. Based on these mean square error values, we observe the partial ordering

$$\mathrm{Med}_{10} \prec G_{10} \prec L_{10}(2) \prec L_{10}(1) \prec T_{10}(2) \tag{5.75}$$

as well as $W_{10}(1) \prec M_{10}(1)$, $W_{10}(2) \prec M_{10}(2)$ and $W_{10}(2) \prec T_{10}(1)$.

Table 5.2

Mean square error of various estimators of μ for n = 10 when a single observation is from $N(\mu+\lambda,1)$ and the others from $N(\mu,1)$

Estimator	λ: 0.0	0.5	1.0	1.5	2.0	3.0	4.0	∞
$\bar{X}_{10}$	0.10000	0.10250	0.11000	0.12250	0.14000	0.19000	0.26000	∞
$T_{10}(1)$	0.10534	0.10791	0.11471	0.12387	0.13285	0.14475	0.14865	0.14942
$T_{10}(2)$	0.11331	0.11603	0.12297	0.13132	0.13848	0.14580	0.14730	0.14745
Med_{10}	0.13833	0.14161	0.14964	0.15852	0.16524	0.17072	0.17146	0.17150
$W_{10}(1)$	0.10437	0.10693	0.11403	0.12405	0.13469	0.15039	0.15627	0.15755
$W_{10}(2)$	0.11133	0.11402	0.12106	0.12995	0.13805	0.14713	0.14926	0.14950
$M_{10}(1)$	0.10432	0.10688	0.11396	0.12385	0.13430	0.14950	0.15513	0.15581
$M_{10}(2)$	0.11125	0.11395	0.12097	0.12974	0.13770	0.14649	0.14853	0.14876
$L_{10}(1)$	0.11371	0.11644	0.12337	0.13169	0.13882	0.14626	0.14797	0.14820
$L_{10}(2)$	0.12097	0.12386	0.13105	0.13933	0.14598	0.15206	0.15310	0.15318
G_{10}	0.12256	0.12549	0.13276	0.14111	0.14777	0.15376	0.15472	0.15479

Realize the dilemma we are in at this juncture as the ordering in (5.75) based on the mean square error is almost a reverse ordering of the estimators given in (5.73) based on the bias. Table 5.2 also reveals that the modified maximum likelihood estimators perform better than the Winsorized means which in turn perform better than the trimmed means when λ is small. This is not surprising since the modified maximum likelihood estimators are almost best linear unbiased estimators (and are also almost the maximum likelihood estimators) based on the n–2r central order statistics. For large values of λ, however, the trimmed mean $T_{10}(2)$ and the modified maximum likelihhod estimator $M_{10}(2)$ both remain optimal. For more details on some properties of these estimators and their applications in developing some robust inference procedures, one could refer to Andrews et al. (1972) and Tiku et al. (1986).

Similarly, for the scale–outlier model we may write the location estimators as

$$M^*(\tau) = \sum_{i=r+1}^{n-r} a_i\, Z^*_{i:n}(\tau)$$

which, when $a_i = a_{n+1-i}$, are clearly symmetrically distributed about 0 for all values of τ. As a result we have $E(M^*(\tau)) = 0$ for all values of τ. Moreover, by comparing equation (5.71) with equation (5.58) we immediately see that the limiting behaviour of $M^*(\tau)$, given $X_n > 0$, as $\tau \to \infty$ corresponds to that of $M(\lambda)$ as $\lambda \to \infty$, and in a similar way the limiting behaviour of $M^*(\tau)$, given $X_n < 0$, as $\tau \to \infty$ corresponds to that of $M(\lambda)$ as $\lambda \to -\infty$. We have, therefore,

$$\lim_{\tau\to\infty} E\left[M^{*2}(\tau)\right] = E\left[M^{*2}(\infty)\right]$$

$$= \Pr(X_n > 0)\, E\left[M^2(\infty)\right] + \Pr(X_n < 0)\, E\left[M^2(-\infty)\right]$$
$$= \frac{1}{2}\left\{E\left[M^2(\infty)\right] + E\left[M^2(-\infty)\right]\right\}$$
$$= E\left[M^2(\infty)\right]. \tag{5.76}$$

Under the scale–outlier model the estimators of location considered earlier are all unbiased for all values of τ. By making use of the table of variances and covariances of normal order statistics under the

Table 5.3

Variance of various estimators of μ for n = 10 when a single observation is from $N(\mu,\tau^2)$ and the others from $N(\mu,1)$

Estimator	τ = 0.5	1.0	2.0	3.0	4.0	∞
$\bar{X}_{10}$	0.09250	0.10000	0.13000	0.18000	0.25000	∞
$T_{10}(1)$	0.09491	0.10534	0.12133	0.12955	0.13417	0.14942
$T_{10}(2)$	0.09953	0.11331	0.12773	0.13389	0.13717	0.14745
Med_{10}	0.11728	0.13833	0.15375	0.15953	0.16249	0.17150
$W_{10}(1)$	0.09571	0.10437	0.12215	0.13221	0.13801	0.15754
$W_{10}(2)$	0.09972	0.11133	0.12664	0.13365	0.13745	0.14950
$M_{10}(1)$	0.09548	0.10432	0.12187	0.13171	0.13735	0.15581
$M_{10}(2)$	0.09940	0.11125	0.12638	0.13328	0.13699	0.14876
$L_{10}(1)$	0.09934	0.11371	0.12815	0.13436	0.13769	0.14820
$L_{10}(2)$	0.10432	0.12097	0.13531	0.14101	0.14398	0.15318
G_{10}	0.10573	0.12256	0.13703	0.14270	0.14565	0.15479

scale–outlier model prepared by David et al. (1977), variance of various estimators of the mean μ of a normal $N(\mu,1)$ population, based on a sample of size n = 10 with one observation being from a normal $N(\mu,\tau^2)$ distribution, have been computed for some specific choices of τ. These values are presented in Table 5.3. From these values we observe that the partial ordering in (5.75) still holds, except for the "inlier" situation $\tau = 0.5$ when $T_{10}(2)$ is inferior to $L_{10}(1)$. We also observe that $W_{10}(1) \prec M_{10}(1)$, $W_{10}(2) \prec M_{10}(2)$, $W_{10}(2) \prec T_{10}(1)$ and $T_{10}(2) \prec M_{10}(2)$ except when τ is very large. These orderings, in general, agree with those based on Table 5.2. As David and Shu (1978) rightly pointed out, this general agreement is less surprising when we remember that there is identity of results not only in the null case ($\lambda = 0$ or $\tau = 1$) but also for the limiting case ($\lambda = \tau = \infty$), the latter by equation (5.76).

Making use of the results given in Section 5.6 and the explicit expressions for the means, variances and covariances of order statistics from a single scale–outlier exponential model, Balakrishnan and Ambagaspitiya (1988) have evaluated the means, variances and covariances of order statistics from a single scale–outlier double exponential model. The model they have considered is that a sample of size n consists of n–1 observations from a Laplace population with pdf

$$f(x) = \frac{1}{2\sigma} e^{-|x-\mu|/\sigma}, \quad -\infty < x < \infty,\ -\infty < \mu < \infty,\ \sigma > 0,$$

while one observation is from a population with pdf

$$g(x) = \frac{1}{2a\sigma} e^{-|x-\mu|/a\sigma}, \quad -\infty < x < \infty, \ \sigma > 0, \ a > 0.$$

After noting that the various estimators of the location parameter μ considered earlier are all unbiased, Balakrishnan and Ambagaspitiya (1988) have studied the performance of these estimators by computing their variance for various choices of a and different sample sizes. The median, the linearly weighted mean and the Gastwirth mean all perform very efficiently in this case and this is to be expected as the double exponential distribution is a symmetric long-tailed distribution. In addition, they have also noted that the trimmed means do better than the corresponding modified maximum likelihood estimators which in turn perform better than the corresponding Winsorized means.

Exercises

1. Suppose $X_1, X_2, \ldots, X_n$ are independent random variables with X_i $(i=1,2,\ldots,n)$ having pdf $f_i(x)$ and cdf $F_i(x)$. Then show that the cdf of $X_{i:n}$ may be written as

$$H_{i:n}(x) = \sum_{r=i}^{n} \sum_{S_r} \prod_{k=1}^{r} F_{j_k}(x) \prod_{k=r+1}^{n} \left\{1-F_{j_k}(x)\right\},$$

where the summation S_r extends over all permutations $(j_1, j_2, \ldots, j_n)$ of $1,2,\ldots,n$ for which $j_1 < j_2 < \ldots < j_r$ and $j_{r+1} < j_{r+2} < \ldots < j_n$.

2. (Sen (1970)). In the above problem, denoting the collection of distribution functions $(F_1, F_2, \ldots, F_n)$ by $\mathscr{F}$, the average cdf by $\bar{F} = \frac{1}{n} \sum_{r=1}^{n} F_r$ and its unique quantile of order p by ξ_p, i.e., $\bar{F}(\xi_p) = p$, show that for $i = 2,3,\ldots,n-1$ and $x \le \xi_{\frac{i-1}{n}} < \xi_{\frac{i}{n}} \le y$,

$$\Pr\left\{x < X_{i:n} \le y \,|\, \mathscr{F}\right\} \ge \Pr\left\{x < X_{i:n} \le y \,|\, \bar{F}\right\},$$

where the equality holds only if $F_1 = F_2 = \ldots = F_n = F$ at both the points x and y. Also show, for all x, that

$$\Pr\left\{X_{1:n} \le x \,|\, \mathscr{F}\right\} \ge \Pr\left\{X_{1:n} \le x \,|\, \bar{F}\right\}$$

and

$$\Pr\left\{X_{n:n} \le x \,|\, \mathscr{F}\right\} \le \Pr\left\{X_{n:n} \le x \,|\, \bar{F}\right\},$$

where the equalities hold only if $F_1 = F_2 = \ldots = F_n = F$ at the point x. (Hint: You may use a result of Hoeffding (1956) on the distribution of successes in n independent trials).

3. As in Problem 1, let $X_{1:n} \le X_{2:n} \le \ldots \le X_{n:n}$ denote the order statistics obtained from n independent random variables X_i $(i = 1,2,\ldots,n)$ having pdf $f_i(x)$ and cdf $F_i(x)$. Then show that:

   (a) the pdf of the largest order statistic $X_{n:n}$ is given by

$$h_{n:n}(x) = \left\{\prod_{r=1}^{n} F_r(x)\right\} \sum_{r=1}^{n} \left\{f_r(x)/F_r(x)\right\};$$

   (b) the cdf of the sample range $W_n = X_{n:n} - X_{1:n}$ is given by

$$\Pr(W_n \le w) = \sum_{r=1}^{n} \int_{-\infty}^{\infty} f_r(x) \prod_{\substack{s=1 \\ s \ne r}}^{n} \{F_s(x+w) - F_s(x)\}\, dx.$$

4. (Cohn et al. (1960)). Suppose X_{ij} $(i = 1,2,\ldots,k;\ j = 1,2,\ldots,n)$ are k independent random samples of size n, with X_{ij} having pdf $f_i(x)$ and cdf $F_i(x)$, $j = 1,2,\ldots,n$. Then show that the maxima from the k samples are the k largest of the kn random variables with probability

$$n^k \int_{-\infty}^{\infty} \left\{\prod_{m=1}^{k} F_m^{n-1}(x)\right\} \sum_{r=1}^{k} \left\{\prod_{\substack{s=1 \\ s \ne r}}^{k} (1-F_s(x))\right\} f_s(x)\, dx.$$

5. (Kelleher and Walsh (1972)). In the sample $X_{1:n} \le X_{2:n} \le \dots \le X_{n:n}$ of ordered continuous variables, suppose $X_{n:n}$ is an outlier in the sense that it arises from a different population than that giving rise to the remainder of the sample. Show then that the confidence coefficient of the interval $(X_{i:n}, X_{n-i+1:n})$, $i=2,3,\dots,[n/2]$, for the median in samples of n still equals $2^{-n} \sum_{r=i}^{n-i} \binom{n}{r}$.

6. (Conover (1965); David (1966)). Suppose k mutually independent random samples of size n, each drawn from a continuous population with cdf F(x), are ordered on the basis of the largest observation in each sample. Further, suppose Y_{ij} $(i = 1,2,\dots,n;\ j = 1,2,\dots,k)$ denote the i'th variate in order of magnitude in the sample whose largest member Y_{1j} has rank j among the k maxima Y_{11}, $Y_{12},\dots,Y_{1k}$. Then show that

$$\Pr(Y_{ij} < x) = \sum_{p=0}^{j-1} \binom{k}{p} \left\{1-F^n(x)\right\}^p \left\{F^n(x)\right\}^{k-p}$$
$$+ \sum_{p=0}^{j-1} \sum_{q=1}^{i-1} \sum_{r=0}^{q-1} j\binom{k}{j}\, q\binom{n}{q} \binom{j-1}{p} \binom{q-1}{r} (-1)^{j+q-r-p} \left[\left\{F(x)\right\}^{n-1-r} - \left\{F(x)\right\}^{nk-np}\right] / \left[nk-np+1-n+r\right],$$

where the triple summation is zero when $i = 1$.

7. (Neyman and Scott (1971)). Let $X_{n:n}$ be called a γ outlier on the right if $X_{n:n} > X_{n-1:n} + \gamma(X_{n-1:n} - X_{1:n})$, $\gamma > 0$. Let us also denote the probability that a sample of n observations from a continuous population with pdf f(x) and cdf F(x) will contain such a γ outlier on the right by $\Pi(\gamma,n,F)$. Then prove that

$$\Pi(\gamma,n,F) = n(n-1)\int_{-\infty}^{\infty} \int_{x}^{\infty} \left\{F\left[\frac{\gamma x+y}{\gamma+1}\right] - F(x)\right\}^{n-2} f(y)\, f(x)\, dy\, dx.$$

Show also that for fixed $n > 2$, the above probability $\Pi(\gamma,n,F)$ is a decreasing function of γ which tends to 0 as $n \to \infty$.

8. (Vaughan and Venables (1972)). Suppose $X_1, X_2,\dots,X_n$ are independently distributed, X_j having density function $f_j(x)$ and cdf $F_j(x)$. Consider a sampling process where one value is taken from each of the above n distributions and the sample ordered so that $X_{1:n} \le X_{2:n} \le \dots \le X_{n:n}$.

a. Show that the pdf of $X_{k_1:n}$, $1 \le k_1 \le n$, is given by

$$h_{k_1:n}(x_1) = \{(k_1-1)!(n-k_1)!\}^{-1} \begin{vmatrix} F_1(x_1) & F_2(x_1) & \cdots & F_n(x_1) \\ \vdots & & & \vdots \\ F_1(x_1) & F_2(x_1) & \cdots & F_n(x_1) \\ f_1(x_1) & f_2(x_1) & \cdots & f_n(x_1) \\ 1-F_1(x_1) & 1-F_2(x_1) & \cdots & 1-F_n(x_1) \\ \vdots & & & \vdots \\ 1-F_1(x_1) & 1-F_2(x_1) & \cdots & 1-F_n(x_1) \end{vmatrix}^{+} \begin{matrix} \left.\right\} k_1-1 \text{ rows} \\ \\ \left.\right\} n-k_1 \text{ rows} \end{matrix}$$

where $|A|^{+}$ denotes the permanent of a square matrix A. The permanent of A is defined like the determinant, except that all signs are positive; for details, see Aitken (1939).

b. Similarly, show that the joint pdf of $X_{k_1:n}$ and $X_{k_2:n}$, $1 \le k_1 < k_2 \le n$, for $x_1 < x_2$, is given by

$$h_{k_1,k_2:n}(x_1,x_2) = \left\{(k_1-1)!(k_2-k_1-1)!(n-k_2)!\right\}^{-1}$$

$$\overset{+}{}\left|\begin{array}{ccc} F_1(x_1) & F_2(x_1) & F_n(x_1) \\ & & \\ F_1(x_1) & F_2(x_1) & F_n(x_1) \\ f_1(x_1) & f_2(x_1) & f_n(x_1) \\ F_1(x_2)-F_1(x_1) & F_2(x_2)-F_2(x_1) & F_n(x_2)-F_n(x_1) \\ & & \\ F_1(x_2)-F_1(x_1) & F_2(x_2)-F_2(x_1) & F_n(x_2)-F_n(x_1) \\ f_1(x_2) & f_2(x_2) & f_n(x_2) \\ 1-F_1(x_2) & 1-F_2(x_2) & 1-F_n(x_2) \\ & & \\ 1-F_1(x_2) & 1-F_2(x_2) & 1-F_n(x_2) \end{array}\right|^{+} \begin{array}{l} \left.\begin{array}{l} \\ \\ \\ \end{array}\right] k_1-1 \text{ rows} \\ \\ \left.\begin{array}{l} \\ \\ \\ \end{array}\right] k_2-k_1-1 \text{ rows} \\ \\ \left.\begin{array}{l} \\ \\ \\ \end{array}\right] n-k_2 \text{ rows.} \end{array}$$

c. Generalize these results and derive the joint density of any subset $X_{k_1:n}, X_{k_2:n}, \ldots, X_{k_p:n}$ of order statistics, where $1 \le k_1 < k_2 < \ldots < k_p \le n$.

d. In particular, show that the density of the full joint distribution of $X_{1:n}, X_{2:n}, \ldots, X_{n:n}$ at $x_1, x_2, \ldots, x_n$ is given by

$$h_{1,2,\ldots,n:n}(x_1,x_2,\ldots,x_n) = \overset{+}{}\left|\begin{array}{ccc} f_1(x_1)\,. & . & .\,f_n(x_1) \\ & & \\ & & \\ f_1(x_n)\,. & . & .\,f_n(x_n) \end{array}\right|^{+}$$

e. For the special case when there is exactly one outlier in the sample, that is, $F_1 = F_2 = \ldots = F_{n-1} = F$ and $F_n = G$, show that the densities in (a) and (b) reduce to the expressions in equations (5.4) and (5.6), respectively.

9. (Balakrishnan (1987b)). For a single–outlier model, show for $n \ge 2$ that

$$\sum_{i=1}^{n} \frac{1}{i}\,\mu_{i:n}^{(k)} = \sum_{i=1}^{n} \frac{1}{i}\,\nu_{i:n}^{(k)} + \frac{1}{n} \sum_{i=1}^{n} \left[\mu_{i:n}^{(k)} - \nu_{i:n}^{(k)}\right]$$

$$= \frac{1}{n} \sum_{i=i}^{n} \mu_{1:i}^{(k)} + \sum_{i=1}^{n-1} \left[\frac{1}{i} - \frac{1}{n}\right] \nu_{1:i}^{(k)}$$

and

$$\sum_{i=1}^{n} \frac{1}{(n-i+1)}\,\mu_{i:n}^{(k)} = \sum_{i=1}^{n} \frac{1}{(n-i+1)}\,\nu_{i:n}^{(k)} + \frac{1}{n} \sum_{i=1}^{n} \left[\mu_{i:i}^{(k)} - \nu_{i:i}^{(k)}\right]$$

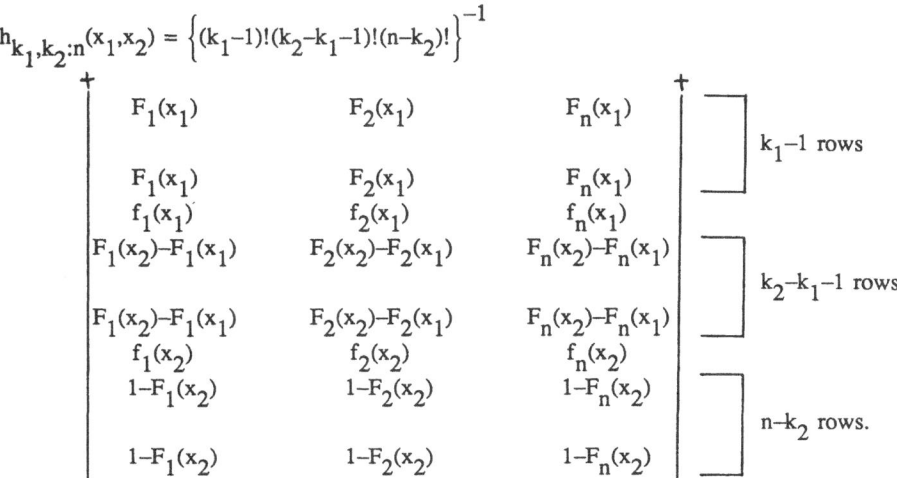

$$= \frac{1}{n} \sum_{i=1}^{n} \mu_{i:i}^{(k)} + \sum_{i=1}^{n-1} \left(\frac{1}{i} - \frac{1}{n}\right) \nu_{i:i}^{(k)}.$$

In particular, by setting $G(x) \equiv F(x)$ in the above results, deduce the identities given in Exercise 4 of Chapter 2.

10. (Balakrishnan (1988a)). For a single–outlier model, show that

$$\sum_{i=1}^{n-1} \mu_{i,i+1:n} + \sum_{j=2}^{n} \binom{n-1}{j-1} \mu_{1,j:j}$$

$$= \sum_{j=1}^{n-1} \left\{\binom{n-1}{j-1} \nu_{1:n-j}\, \mu_{j:j} + \binom{n-1}{j} \nu_{j:j}\, \mu_{1:n-j}\right\} - \sum_{j=2}^{n-1} \binom{n-1}{j} \nu_{1,j:j}.$$

11. Assuming $X_1, X_2, \ldots, X_{n-1}$ to be a random sample from a standard exponential distribution with pdf $f(x) = e^{-x}$, $0 \le x < \infty$, and Y to be an independent random variable with pdf $g(x) = \frac{1}{\sigma} e^{-x/\sigma}$, $x \ge 0$, $\sigma > 0$, and then denoting $Z_{1:n} \le Z_{2:n} \le \ldots \le Z_{n:n}$ for the order statistics obtained from these n variables, derive $E(Z_{i:n})$, $Var(Z_{i:n})$ and $Cov(Z_{i:n}, Z_{j:n})$.

12. Assuming Y to be an independent random variable with pdf $g(x) = e^{-(x-\mu)}$, $x \ge \mu$, once again derive $E(Z_{i:n})$, $Var(Z_{i:n})$ and $Cov(Z_{i:n}, Z_{j:n})$.

13. (Balakrishnan and Ambagaspitiya (1988)). Making use of the expressions of means, variances and covariances of order statistics derived in Exercise 11 in Relations 5.21 and 5.22, compute the means, variances and covariances of order statistics from a single scale–outlier double exponential model, viz.,

$$X_1, \ldots, X_{n-1} \text{ with pdf } \frac{1}{2} e^{-|x|}, \quad -\infty < x < \infty$$

and

$$Y \text{ with pdf } \frac{1}{2\sigma} e^{-|x|/\sigma}, \quad -\infty < x < \infty,\ \sigma > 0,$$

for $n = 10$ and 20, and $\sigma = 0.5, 1.0(1.0)10.0$. Making use of these values, compare the efficiency of the various estimators of the location parameter considered in Section 8.

14. (Smith and Tong (1983); David (1986)). Let $x_r = y_r + z_r$, $r = 1,2,\ldots,n$, and write $x_{(1)} \le x_{(2)} \le \ldots \le x_{(n)}$ or $x_{[1]} \ge x_{[2]} \ge \ldots \ge x_{[n]}$, and likewise for y_r and z_r. Then, show for $1 \le i \le n$, that

$$x_{[i]} \le \min_{r=1,2,\ldots,i} \left(y_{[r]} + z_{[i+1-r]}\right)$$

and

$$x_{(i)} \ge \max_{r=1,2,\ldots,i} \left(y_{(r)} + z_{(i+1-r)}\right).$$

Thence, derive the inequality

$$E\left[X_{(i)}\right] \geq \max_{r=1,2,\ldots i} \left[E(Y_{(r)}) + E(Z_{(i+1-r)})\right].$$

15. (David (1986)). For a single–location outlier model, defining $Z_r = \delta > 0$ for some unknown value of r and $Z_r = 0$ otherwise, making use of the results in Exercise 14 show that for $i = 1, 2, \ldots, n-1$

$$E\left[Y_{(i)}\right] \leq E\left[X_{(i)}\right] \leq \min\left[E(Y_{(i+1)}), E(Y_{(i)}) + \delta\right],$$

and for $i = n$

$$\max\left[E(Y_{(n)}), E(Y_{(1)}) + \delta\right] \leq E(X_{(n)}) \leq E(Y_{(n)}) + \delta.$$

16. (Smith and Tong (1983); David (1986)). Let $x'_{(i)}$, $i=1,2,\ldots,n$, denote the n sums $y_{(i)} + z_{(n+1-i)}$ arranged in increasing order of magnitude. Further, let ℓ be a convex linear function of ordered x_i, viz. $\ell = \sum_{i=1}^{n} C_i x_{(i)}$ with $C_1 \leq C_2 \leq \ldots \leq C_n$. Then show that for $C_1 \geq 0$

$$\sum_{i=1}^{n} C_i x'_{(i)} \leq \ell \leq \sum_{i=1}^{n} C_i \left[y_{(i)} + z_{(i)}\right].$$

Also, show that the above inequality continues to hold if $C_m < 0$, $C_{m+1} \geq 0$, $m = 1,2,\ldots,n$.

17. Using the notation in Section 5.1, without assuming absolute continuity verify that

$$i\, H_{i+1:n}(x) + (n-i)\, H_{i:n}(x) = (n-1)\, H_{i:n-1}(x) + F_{i:n-1}(x)$$

for $1 \leq i \leq n-1$. [Hint: Use equations (5.3) and (2.68)]. Due to the above result we may note that Relation 5.2 continues to hold without the assumption of absolute continuity.

18. Denoting the joint c.d.f. of $Z_{r:n}$ and $Z_{s:n}$ $(1 \leq r < s \leq n)$ by $H_{r,s:n}(x,y)$, $x \leq y$, show that

$$(i-1)H_{i,j:n}(x,y) + (j-i)H_{i-1,j:n}(x,y) + (n-j+1)H_{i-1,j-1:n}(x,y) = (n-1)H_{i-1,j-1:n-1}(x,y) + F_{i-1,j-1:n-1}(x,y)$$

for $2 \leq i < j \leq n$. Due to this result we may note that Relation 5.9 continues to hold without the assumption of absolute continuity.

19. Let $X_1, X_2, \ldots, X_n$ be i.i.d. from an arbitrary distribution function F. Let $U_1, U_2, \ldots, U_n$ be i.i.d. Uniform $(0,\delta)$, independent of the X's. For $i = 1,2,\ldots,n$, define $Y_i = X_i + U_i$. Prove that for $1 \leq i \leq n$, $Y_{i:n} \overset{d}{\to} X_{i:n}$ as $\delta \to 0$. In addition, if the X_i's have a finite k'th moment, show that $E(Y^k_{i:n}) \to E(X^k_{i:n})$ as $\delta \to 0$. Since the Y_i's are absolutely connot refer to densities continue to hold for general distributions.

20. (Balakrishnan (1988c)). (i) Let $X_1, X_2, \ldots, X_n$ be independent variables with X_i $(i = 1,2,\ldots,n)$ having pdf $f_i(x)$ and cdf $F_i(x)$. Let $X_{1:n} \leq X_{2:n} \leq \ldots \leq X_{n:n}$ be the order statistics obtained from the realizations of $X_1, X_2, \ldots, X_n$. Then the density of $X_{i:n}$ $(1 \leq i \leq n)$, viz., $h_{i:n}(x)$, and the joint density of $X_{i:n}$ and $X_{j:n}$ $(1 \leq i < j \leq n)$, viz., $h_{i,j:n}(x,y)$, are as given in Exercise 8.
Now let us denote $h^{[r_1, r_2, \ldots, r_m]}_{i:n-m}(x)$, $1 \leq i \leq n-m$, for the density function of the i'th order statistic

in a sample of size n–m obtained by dropping $X_{r_1}, X_{r_2}, \ldots, X_{r_m}$ from the original set of n variables; similarly, let us denote $h^{[r]}_{i,j:n-1}(x,y)$, $1 \le i < j \le n-1$, for the joint density of the i'th and j'th order statistics in a sample of size n–1 obtained by dropping X_r from the original set of n variables. Then show that for $1 \le i \le n-1$,

$$ih_{i+1:n}(x) + (n-i)h_{i:n}(x) = \sum_{r=1}^{n} h^{[r]}_{i:n-1}(x)$$

and for $2 \le i < j \le n$,

$$(i-1)\, h_{i,j:n}(x,y) + (j-i)h_{i-1,j:n}(x,y) + (n-j+1)h_{i-1,j-1:n}(x,y)$$
$$= \sum_{r=1}^{n} h^{[r]}_{i-1,j-1:n-1}(x,y).$$

(ii) For the p–outlier model, that is, $F_1 = F_2 = \ldots = F_{n-p} = F$ and $F_{n-p+1} = \ldots = F_n = G$, then deduce the relations

$$ih_{i+1:n}(x) + (n-i)h_{i:n}(x) = (n-p)h^{[F]}_{i:n-1}(x) + ph^{[G]}_{i:n-1}(x),\ 1 \le i \le n-1,$$

and

$$(i-1)h_{i,j:n}(x,y) + (j-i)h_{i-1,j:n}(x,y) + (n-j+1)h_{i-1,j-1:n}(x,y)$$
$$= (n-p)h^{[F]}_{i-1,j-1:n-1}(x,y) + ph^{[G]}_{i-1,j-1:n-1}(x,y),\ 2 \le i < j \le n,$$

where $h^{[F]}_{i:n-1}(x)$ and $h^{[G]}_{i:n-1}(x)$ are the density functions of the i'th order statistic in a sample of size n–1 from the p–outlier model and the (p–1) – outlier model, respectively, and similarly for the joint density functions.

(iii) Denote

$$S_{1:n-m}(x) = \sum_{1 \le r_1 < r_2 < \ldots < r_m \le n} h^{[r_1, r_2, \ldots, r_m]}_{1:n-m}(x)$$

and

$$S_{n-m:n-m}(x) = \sum_{1 \le r_1 < r_2 < \ldots < r_m \le n} h^{[r_1, r_2, \ldots, r_m]}_{n-m:n-m}(x)$$

with $S_{1:n}(x) \equiv h_{1:n}(x)$ and $S_{n:n}(x) \equiv h_{n:n}(x)$. Then by repeated application of the first relation in (i), show that

$$h_{i:n}(x) = \sum_{j=i}^{n} (-1)^{j-i} \binom{j-1}{i-1} S_{j:j}(x),\ 1 \le i \le n-1,$$

and

$$h_{i:n}(x) = \sum_{j=n-i+1}^{n} (-1)^{j-n+i-1} \binom{j-1}{n-i} S_{1:j}(x),\ 2 \le i \le n$$

21. (Balakrishnan, 1989a). (i) Let $X_1, X_2, \ldots, X_n$ be independent random variables with X_i $(i = 1,2,\ldots,n)$ having pdf $f_i(x)$ and cdf $F_i(x)$. Then by making use of the relations given in Exercise 20, show that

$$(i-1)\, \sigma_{i,j:n} + (j-i)\, \sigma_{i-1,j:n} + (n-j+1)\, \sigma_{i-1,j-1:n}$$

$$= \sum_{r=1}^{n} \left\{ \sigma^{[r]}_{i-1,j-1:n-1} + \left[\mu^{[r]}_{i-1:n-1} - \mu_{i-1:n}\right] \left[\mu^{[r]}_{j-1:n-1} - \mu_{j:n}\right] \right\}$$

for $2 \le i < j \le n$, where $\sigma_{i,j:n}$ and $\sigma^{[r]}_{i,j:n-1}$ denote the covariances of the i'th and j'th order statistics in samples of size n and n–1 (with X_r dropped), respectively.

(ii) For the p–outlier model, that is, $F_1 = F_2 = \ldots = F_{n-p} = F$ and $F_{n-p+1} = \ldots = F_n = G$, deduce the relation

$$(i-1)\,\sigma_{i,j:n} + (j-i)\,\sigma_{i-1,j:n} + (n-j+1)\,\sigma_{i-1,j-1:n}$$

$$= (n-p) \left\{ \sigma^{[F]}_{i-1,j-1:n-1} + \left[\mu^{[F]}_{i-1:n-1} - \mu_{i-1:n}\right] \left[\mu_{j-1:n-1} - \mu_{j:n}\right] \right\}$$

$$+ p \left\{ \sigma^{[G]}_{i-1,j-1:n-1} + \left[\mu^{[G]}_{i-1:n-1} - \mu_{i-1:n}\right] \left[\mu^{[G]}_{j-1:n-1} - \mu_{j:n}\right] \right\},$$

where, as before, [F] and [G] denote the quantities in a sample of size n–1 from the p–outlier model and the (p–1)–outlier model, respectively.

22. (Balakrishnan, 1989b). Let $X_{1:n} \le X_{2:n} \le \ldots \le X_{n:n}$ denote the order statistics obtained from n independent absolutely continuous random variables X_i $(i = 1,2,\ldots,n)$, with X_i having pdf $f_i(x)$ and cdf $F_i(x)$. Let us denote the single and product moments of these order statistics by $\mu^{(k)}_{i:n}$ $(1 \le i \le n;\ k \ge 1)$ and $\mu_{i,j:n}$ $(1 \le i < j \le n)$.

Let the density functions $f_i(x)$ be all symmetric about 0. Then for $x > 0$, let

$G_i(x) = 2\,F_i(x) - 1$ and $g_i(x) = 2\,f_i(x)$;

that is, the density functions $g_i(x)$, $i = 1,2,\ldots,n$, are obtained by folding the density functions $f_i(x)$ at zero. Let $Y_{1:n} \le Y_{2:n} \le \ldots \le Y_{n:n}$ denote the order statistics obtained from n independent absolutely continuous random variables Y_i $(i = 1,2,\ldots,n)$, with Y_i having pdf $g_i(x)$ and cdf $G_i(x)$. Further, let us denote $\nu^{(k)[r_1,\ldots,r_\ell]}_{i:n-\ell}$ for the k'th single moment of $Y^{[r_1,\ldots,r_\ell]}_{i:n-\ell}$ and $\nu^{[r_1,\ldots,r_\ell]}_{i,j:n-\ell}$ for the product moment of $Y^{[r_1,\ldots,r_\ell]}_{i:n-\ell}$ and $Y^{[r_1,\ldots,r_\ell]}_{j:n-\ell}$, where $Y^{[r_1,\ldots,r_\ell]}_{i:n-\ell}$ denotes the i'th order statistic in a sample of size $n-\ell$ obtained by dropping $Y_{r_1}, Y_{r_2}, \ldots, Y_{r_\ell}$ from the original set of n variables $Y_1, Y_2, \ldots, Y_n$.

Then show that:

(i) For $1 \le i \le n$ and $k = 1,2,\ldots,$

$$\mu^{(k)}_{i:n} = 2^{-n} \left\{ \sum_{\ell=0}^{i-1} \sum_{1 \le r_1 < \ldots < r_\ell \le n} \nu^{(k)[r_1,\ldots,r_\ell]}_{i-\ell:n-\ell} + (-1)^k \sum_{\ell=i}^{n} \sum_{1 \le r_1 < \ldots < r_{n-\ell} \le n} \nu^{(k)[r_1,\ldots,r_{n-\ell}]}_{\ell-i+1:\ell} \right\};$$

(ii) For $1 \le i < j \le n$,

$$\mu_{i,j:n} = 2^{-n} \left\{ \sum_{\ell=0}^{i-1} \sum_{1 \le r_1 < \ldots < r_\ell \le n} \nu^{[r_1,\ldots,r_\ell]}_{i-\ell,j-\ell:n-\ell} + \sum_{\ell=j}^{n} \sum_{1 \le r_1 < \ldots < r_{n-\ell} \le n} \nu^{[r_1,\ldots,r_{n-\ell}]}_{\ell-j+1,\ell-i+1:\ell} \right.$$

$$\left. - \sum_{\ell=i}^{j-1} \sum_{1 \le r_1 < \ldots < r_\ell \le n} \nu^{[r_1,\ldots,r_\ell]}_{j-\ell:n-\ell}\ \nu^{[r_{\ell+1},\ldots,r_n]}_{\ell-i+1:\ell} \right\}.$$

(iii) Verify that for a single–outlier model, these results reduce to Relations 5.21 and 5.22.

23. (Balakrishnan, 1989c). Let $X_1, X_2, \ldots, X_n$ be independent random variables with X_i $(i = 1,2,\ldots,n)$ having pdf $f_i(x)$ and cdf $F_i(x)$. Then by making use of the relations given in Exercise 20, show that:

(i) for $1 \le i \le m \le n-1$ and $k = 1,2,\ldots,$

$$\sum_{1 \le r_1 < \ldots < r_{n-m} \le n} \mu_{i:m}^{(k)[r_1,\ldots,r_{n-m}]} = \sum_{r=0}^{n-m} \binom{i-1+r}{r} \binom{n-i-r}{n-m-r} \mu_{i+r:n}^{(k)},$$

and

(ii) for $1 \le i < j \le m \le n-1$,

$$\sum_{1 \le r_1 < \ldots < r_{n-m} \le n} \mu_{i,j:m}^{[r_1,\ldots,r_{n-m}]} = \sum_{r=0}^{n-m} \sum_{s=r}^{n-m} \binom{i-1+r}{r} \binom{j-i-1+s-r}{s-r} \binom{n-j-s}{n-m-s} \mu_{i+r,s+j:n}.$$

For the p–outlier model, that is, $F_1 = \ldots = F_{n-p} = F$ and $F_{n-p+1} = \ldots = F_n = G$, by denoting $\mu_{i:m}^{(k)}[r]$ and $\mu_{i,j:m}[r]$ for the single and the product moments of order statistics in a sample of size m with r outliers present in it, deduce the following relations:

(iii) for $0 \le p \le n$ and $1 \le i \le m \le n-1$,

$$\sum_{r=0}^{p} \binom{n-p}{n-m-r} \binom{p}{r} \mu_{i:m}^{(k)}[p-r] = \sum_{r=0}^{n-m} \binom{i-1+r}{r} \binom{n-i-r}{n-m-r} \mu_{i+r:n}^{(k)}[p]$$

and

(iv) for $0 \le p \le n$ and $1 \le i < j \le m \le n-1$,

$$\sum_{r=0}^{p} \binom{n-p}{n-m-r} \binom{p}{r} \mu_{i,j:m}[p-r] = \sum_{r=0}^{n-m} \sum_{s=r}^{n-m} \binom{i-1+r}{r} \binom{j-i-1+s-r}{s-r} \binom{n-j-s}{n-m-s} \mu_{i+r,s+j:n}[p].$$

CHAPTER 6

RECORD VALUES

6.0. Introduction

Chandler (1952) was the first to consider record values in the context of weather conditions. Subsequently the indirect relationship between record values and order statistics was recognized and exploited. It is consequently not surprising that many of the results of Chapters 1 through 4 have parallels relating to expectations of record values. We begin by defining the concept of record values.

6.1. Record values

Let $X_1, X_2, \ldots$ be a sequence of independent identically distributed random variables with common distribution function F. Although some results are available for discrete distributions, the major corpus of theory deals with the case where F is a continuous distribution function. Attention in the present chapter will be restricted to such a setting. An observation X_j is a record (an upper record) if $X_j > X_i$ for every i < j. Record times $\{L_n\}_{n=0}^{\infty}$ are defined by

$$L_0 = 1$$

and (6.1)

$$L_n = \min\{j: j > L_{n-1}, X_j > X_{L_{n-1}}\}.$$

The record value sequence $\{Y_n\}_{n=0}^{\infty}$ is then defined by

$$Y_n = X_{L_n}, \quad n = 0,1,2,\ldots . \tag{6.2}$$

A particularly simple case involves exponentially distributed random variables. Utilizing the lack of memory property of exponential random variables it is evident that the differences between successive records will be themselves i.i.d. exponential random variables. In a sense, the unit exponential distribution is associated with a canonical record value sequence. Let us denote a sequence of i.i.d. exponential (1) random variables by $\{X_i^*\}_{i=1}^{\infty}$ and the corresponding record value sequence by $\{Y_n^*\}_{n=0}^{\infty}$. As remarked above

$$Y_n^* - Y_{n-1}^* \text{ are i.i.d. } \exp(1) \tag{6.3}$$

and consequently

$$Y_n^* \sim \Gamma(n + 1, 1), \quad n = 0,1,2,\ldots. \tag{6.4}$$

Now if $X_n \sim F$, a continuous distribution, then clearly

$$X_n \overset{d}{=} F^{-1}(1 - e^{-X_n^*}) \tag{6.5}$$

where $X_n^* \sim \exp(1)$, equivalently

$$-\log \bar{F}(X_n) \sim \exp(1). \tag{6.6}$$

Since $F^{-1}(1 - e^{-x})$ is a monotone function it is evident that the record value sequences of X_n and X_n^* are related in a manner dictated by (6.5). Thus

$$Y_n \overset{d}{=} F^{-1}(1 - e^{-Y_n^*}) \tag{6.7}$$

where $\{Y_n^*\}$ is the record value sequence of a sequence of i.i.d. unit exponential variates. The known distribution of $Y_n^*(\sim \Gamma(n+1,1))$ allows us to immediately, via (6.7), write the following expression for the d.f. of Y_n

$$P(Y_n > y) = \bar{F}(y) \sum_{k=0}^{n} [-\log \bar{F}(y)]^k/k!. \tag{6.8}$$

If the distribution F is absolutely continuous with density f then (6.8) can be differentiated to yield

$$f_{Y_n}(y) = f(y)[-\log \bar{F}(y)]^n/n!\ . \tag{6.9}$$

If we recall the central limit theorem we see that for large n

$$Y_n \overset{\cdot}{\sim} F^{-1}(1 - e^{-nz}) \tag{6.10}$$

(in which $\overset{\cdot}{\sim}$ denotes "approximately distributed") where $Z \sim N(1,1)$. Observation (5.10) can be parlayed into an interpretation of the possible classes of limit distributions for Y_n. Resnick (1973) in fact shows that the three possible limit laws for $(Y_n^* - b_n)/a_n$ are

$$\Phi(x), \tag{6.11}$$

$$\Phi_{1,\alpha}(x) = \Phi(\log x^{\alpha}), \quad x \geq 0, \tag{6.12}$$

and

$$\Phi_{2,\alpha}(x) = \Phi(\log (-x)^{-\alpha}), \quad x < 0 \tag{6.13}$$

where $\Phi(x)$ is the standard normal distribution. A clever duality theorem (Resnick's Theorem 4.1) relates the domain of attraction of the record value Y_n to the domain of attraction of maxima from a sequence of i.i.d. observations from the "associated" distribution

$$H(x) = 1 - \exp\left[-\sqrt{-\log \bar{F}(x)}\,\right]. \tag{6.14}$$

Necessary and sufficient conditions for the existence of $E(Y_n)$ were discussed by Nagaraja (1978). The simplest sufficient condition (proved using (6.8) and the Holder inequality) for the existence of $E(|Y_n|)$ for all n is the existence of $E|X_i|^p$ for some $p > 1$. For convenience we will assume this regularity condition henceforth.

Before moving on to discuss bounds on expected records, it is convenient to discuss another parent distribution, the uniform (0,1), whose record value sequence is readily analysed. Suppose $\{\tilde{X}_n\}_{n=1}^{\infty}$ are i.i.d. uniform (0,1) random variables. Denote the corresponding sequence of records by $\{\tilde{Y}_n\}_{n=0}^{\infty}$. Rather than use (6.8), let us determine the distribution of $\tilde{Y}$ directly. Given $\tilde{Y}_n = y$, say, it is clear that the next record $\tilde{Y}_{n+1}$ will be uniformly distributed over the interval (y,1). Alternatively we may write

$$(1 - \tilde{Y}_{n+1}) \stackrel{d}{=} U(1 - \tilde{Y}_n) \tag{6.15}$$

where U ~ uniform (0,1) is independent of $\tilde{Y}_n$. Since $1 - \tilde{Y}_0$ itself is uniform (0,1), (6.15) yields the convenient representation

$$(1 - \tilde{Y}_n) \stackrel{d}{=} \prod_{i=0}^{n} U_i \tag{6.16}$$

where $\{U_i\}_{i=0}^{\infty}$ are i.i.d. uniform (0,1) random variables. A straightforward inductive argument using (6.16) yields the density for $\tilde{Y}_n$ in the form

$$f_{\tilde{Y}_n}(y) = [-\log(1-y)]^n/n!, \quad 0 < y < 1 \tag{6.17}$$

(see Exercise 2). The same result can be obtained if we substitute $\bar{F}(y) = 1 - y$ in (6.9). A sequence of record values corresponding to an arbitrary distribution F can be studied with reference to the record value sequence for a uniform sequence. If $\tilde{X}$ is uniform (0,1) then $F^{-1}(\tilde{X})$ has distribution F and the record value sequence $\{Y_n\}$ corresponding to F satisfies

$$Y_n = F^{-1}(\tilde{Y}_n). \tag{6.18}$$

Following Nagaraja (1978) we can exploit the relation (6.18) to determine bounds on $E(Y_n)$ as we shall see in the following section.

Naturally one can define lower records in a manner analogous to that used for upper records. The simplest way to do this is as follows. Define $X_n' = -X_n$ and let Y_n' be the sequence of upper record values of X_n'. The lower record value sequence $\{Y_n''\}_{n=0}^{\infty}$ corresponding to $\{X_n\}$ is defined by

$$Y_n'' = -Y_n' . \tag{6.19}$$

The corresponding density and distribution functions are

$$f_{Y_n''}(y) = f(y)[-\log F(y)]^n/n! \tag{6.20}$$

and

$$F_{Y_n''}(y) = F(y) \sum_{k=0}^{n} [-\log F(y)]^k/k!. \tag{6.21}$$

These are readily obtained from (6.8) and (6.9) or alternatively by first considering the simple case of a lower record value sequence corresponding to uniform (0,1) X_i's (see Exercise 3).

Houchens (1984) discussed the joint distribution of upper and lower records. It is a relatively simple computation since obviously $Y_0 = Y_0''$ and for j,k ≥ 2, Y_j'' and Y_k are conditionally independent given $Y_0(=X_1)$. We find for $y'' < y$

$$f_{Y_j'',Y_k}(y'',y) = \frac{f(y'')f(y)}{(j-1)!(k-1)!} \int_{F(y'')}^{F(y)} \frac{\{\log(u/F(y''))\}^{j-1}\{\log((1-u)/(1-F(y)))\}^{k-1}}{u(1-u)}. \tag{6.22}$$

The difference between upper and lower records in general does not seem to have a tractable distribution. It is not even clear that it is of any practical interest. In contrast the record range sequence has some potential interest. It is defined as follows. Consider the sequence of sample ranges $X_{n:n} - X_{1:n}$, n = 1,2,... . Define $R_0 = 0$ and R_i (i = 1,2,...) to be the i'th record in the sequence of sample ranges. A record range is encountered whenever either a new lower or upper record is observed. Using the computations described in Exercise 4 one finds for i = 1,2,...

$$f_{R_i}(r) = \frac{2^i}{(i-1)!} \int_{-\infty}^{\infty} f(r+z)f(z)[-\log(1 - F(z+r) + F(z))]^{i-1}\,dz. \tag{6.23}$$

The only case in which further simplification is possible is the case in which the X_i's are i.i.d. uniform random variables. If $X_i = \tilde{X}_i \sim$ uniform (0,1) then the corresponding i'th record range $\tilde{R}_i$ has its density given by

$$f_{\tilde{R}_i}(r) = \frac{2^i(1-r)}{(i-1)!}[-\log(1-r)]^{i-1}, \quad 0 < r < 1. \tag{6.24}$$

Curiously this i'th record range has the same distribution as the (i–1)'th record value in a sequence of i.i.d. random variables each distributed like the minimum of two independent uniform (0,1) random variables. There should be a heuristic explanation of this phenomenon although none has been supplied.

6.2. Bounds on mean record values

Much of this section is taken from Nagaraja (1978). Let Y_n denote the n'th (upper) record corresponding to a sequence $X_1, X_2, \ldots$ of random variables which are i.i.d. F. Assume that $E(X_i^2) < \infty$ and consequently without loss of generality that $E(X_i) = 0$ and $\text{var}(X_i) = 1$. Using (6.17) and (6.18) we may write

$$E(Y_n) = \int_0^1 F^{-1}(u)[-\log(1-u)]^n/n!\,du. \tag{6.25}$$

We wish to maximize (6.25) subject to $\int_0^1 F^{-1}(u)du = 0$ and $\int_0^1 [F^{-1}(u)]^2 du = 1$. The Schwarz inequality can be used here exactly as it was in Chapter 3. Define

$$g(u) = \{[-\log(1-u)]^n/n!\} - 1, \quad 0 < u < 1, \tag{6.26}$$

then we see that

$$E(Y_n) = \int_0^1 F^{-1}(u)g(u)du \le \sqrt{\int_0^1 g^2(u)du}$$

since $\int_0^1 (F^{-1})^2 du = 1$, with equality iff $F^{-1}(u) \propto g(u)$ on (0,1). It is not difficult to integrate $g^2(u)$. We thus obtain

$$E(Y_n) \le \sqrt{\binom{2n}{n} - 1}. \tag{6.27}$$

The extremal distribution is readily shown to be that of translated Weibull random variable. Specifically let Z_i have as its survival function

$$P(Z_i > z) = \exp(-z^{1/n}), \quad z > 0$$

and define

$$X_i = [(2n)! - (n!)^2]^{-1/2} Z - \left[\binom{2n}{n} - 1\right]^{-1/2} \tag{6.28}$$

then $F_{X_i}^{-1}(u) \propto g^2(u)$ with $E(X_i) = 0$, $E(X_i^2) = 1$ and so for such a sequence $\{X_i\}$ equality is achieved in (6.25).

If the parent distribution F is assumed to be symmetric with mean 0 and variance 1 we may write

$$E(Y_n) = \frac{1}{n!}\int_{1/2}^{1} F^{-1}(u)\{[-\log(1-u)]^n - (-\log u)^n\}du. \tag{6.29}$$

Applying the Schwarz inequality we get for any symmetric F with mean 0 and variance 1,

$$E(Y_n) \le \frac{1}{\sqrt{2}}\left\{\binom{2n}{n} - \frac{1}{(n!)^2}\int_0^1 [\log u \log(1-u)]^n\,du\right\}^{1/2}. \tag{6.30}$$

Nagaraja (1978) supplies a short table of these bounds. For large n, the ratio of the bounds (6.27) and (6.30) is approximately $\sqrt{2}$.

As in Section 1 of Chapter 3, we can obtain tighter, albeit more complicated, bounds using a convenient orthonormal series. If we use the Legendre polynomials $\{\psi_i\}_{i=0}^{\infty}$ (defined in (3.24)) we get

(i) for an arbitrary parent distribution with mean 0 and variance 1,

$$E(Y_n) \le \sum_{k=1}^{m} a_k b_k + \sqrt{1 - \sum_{k=1}^{m} a_k^2}\left\{\binom{2n}{n} - 1 - \sum_{k=1}^{m} b_k^2\right\}^{1/2} \tag{6.31}$$

where

$$a_k = \int_0^1 F^{-1}(u)\psi_k(u)du$$

and

$$b_k = \int_0^1 \{[-\log(1-u)]^n/n!\}\,\psi_k(u)du$$

and

(ii) for a symmetric parent distribution with mean 0 and variance 1,

$$E(Y_n) \le \sum_{k=1}^{m} a_{2k+1} b_{2k+1}$$

$$+ \sqrt{1 - \sum_{k=1}^{m} a_{2k+1}^2}\left\{\frac{(2n)! - \int_0^1 [\log u \ \log(1-u)]^n du}{2(n!)^2} - \sum_{k=0}^{m} b_{2k+1}^2\right\}^{1/2}. \tag{6.32}$$

If $\{Y_n\}$ is a sequence of record values corresponding to a parent distribution F while $\{Y_n^*\}$ is a sequence corresponding to F^*, it is sometimes possible to order $F(E(Y_n))$ and $F^*(E(Y_n^*))$ using Jensen's inequality and the representation (6.25). Specifically if $F^{*-1}F$ is convex on the support of F then

$$F(E(Y_n)) \le F^*(E(Y_n^*)). \tag{6.33}$$

In particular if F^* is IFR then we can apply (6.33) with $F(x) = 1 - e^{-x}$. In this case we know from (6.4) that $E(Y_n) = (n + 1)$ so we get, for F^* an IFR distribution,

$$F^*(E(Y_n)) \ge 1 - e^{-(n+1)}. \tag{6.34}$$

Convex comparisons with symmetric distributions are also possible. Suppose F^* and F are symmetric about 0 and that $F^{*-1}F$ is convex on the positive support of F. It follows that

$$F(E(Y_n)) \le F^*(E(Y_n^*)). \tag{6.35}$$

If in (6.35) we let F^* be symmetric and unimodal and F be the uniform (–1,1) distribution we conclude that

$$F^*(E(Y_n^*)) \ge 1 - 2^{-(n+1)}. \tag{6.36}$$

The inequality in (6.36) is reversed if F^* is symmetric and U–shaped.

6.3. Record values in dependent sequences

Instead of having the basic sequence $\{X_n\}_{n=1}^{\infty}$ be i.i.d. random variables, it is interesting to explore the effects of various types of dependence.

If the X_n's form an exchangeable sequence with common marginal distribution F then, by the de Finetti theorem, $\{X_n\}$ is distributed as a mixture of i.i.d. sequences. Thus there exists a real random variable Z and a family of distributions $\{F_z(x): z \in \mathbb{R}\}$ such that for every n

$$F_{X_1,\dots,X_n}(x_1,\dots,x_n) = \int_{-\infty}^{\infty} \left\{\prod_{i=1}^{n} F_z(x_i)\right\} dF_Z(z). \tag{6.37}$$

Since the marginal distribution of the X's is F we must have

$$F(x) = \int_{-\infty}^{\infty} F_z(x)dF_Z(z). \tag{6.38}$$

The representation (6.37) permits us to compute the conditional distribution of the n'th record Y_n given Z using the results of Section 2. Thus we get (using (6.8))

$$\bar{F}_{Y_n}(y) = \int_{-\infty}^{\infty} \left\{\bar{F}_z(y) \sum_{k=0}^{n} [-\log \bar{F}_z(y)]^k/k!\right\} dF_Z(z) \tag{6.39}$$

and assuming densities, using (6.9),

$$f_{Y_n}(y) = \int_{-\infty}^{\infty} \{f_z(y)[-\log \bar{F}_z(y)]^n/n!\}\, dF_Z(z). \tag{6.40}$$

The expected records are obtainable using (6.25) to evaluate $E(Y_n|Z=z)$. Thus

$$E(Y_n) = \int_{-\infty}^{\infty} \int_0^1 F_z^{-1}(u)[-\log(1-u)]^n/n!\; du\; dF_Z(z).$$

Provided $E|X_1|^p < \infty$ for some $p > 1$, we can interchange the order of integration obtaining

$$E(Y_n) = \int_0^1 \tilde{F}^{-1}(u)[-\log(1-u)]^n/n!\; du \tag{6.41}$$

where

$$\tilde{F}^{-1}(u) = \int_{-\infty}^{\infty} F_z^{-1}(u)dF_Z(z). \tag{6.42}$$

It will be observed that (6.41) has the same form as the corresponding expression for $E(Y_n)$ in the i.i.d. case (Equation (6.25)). However the function $\tilde{F}^{-1}$ appropriate for the exchangeable sequence is in general distinct from F^{-1} (where F is given by (6.38)).

We will illustrate these considerations in the case where $\{X_n\}$ is a particular exchangeable sequence with Pareto distributed marginals. Specifically assume that $Z \sim \Gamma(2,1)$ (i.e. $f_Z(z) = ze^{-z}$, $z > 0$) and given $Z = z$, the X_i's are i.i.d. exp(z) (i.e. $P(X_i > x | Z = z) = e^{-zx}$). It is evident that

$$P(X_i > x) = \int_0^\infty e^{-zx}\, ze^{-z}\, dz = (1 + x)^{-2}, \quad x > 0 \tag{6.43}$$

so that indeed the X_i's are Pareto distributed. In this case for each $z > 0$,

$$F_z(x) = 1 - e^{-zx} \tag{6.44}$$

so that

$$F_z^{-1}(u) = -\frac{1}{z}\log(1-u). \tag{6.45}$$

The expected n'th record for this exchangeable process will be given by (6.41) with $\tilde{F}^{-1}$ defined by (6.42). In particular for the present process

$$\tilde{F}^{-1}(u) = \int_0^\infty -\frac{1}{z}\log(1-u)\, ze^{-z}\, dz = -\log(1-u). \tag{6.46}$$

(note that this is the inverse distribution of a standard exponential distribution, it is not the same as F^{-1} where F is the Pareto distribution given by (6.43)). Substituting (6.46) in (6.41) and integrating we get

$$E(Y_n) = n + 1. \tag{6.47}$$

Thus the exchangeable sequence with $Z \sim \Gamma(2,1)$ has exactly the same sequence of expected record values as does the i.i.d. exponential (1) sequence. Note that this does not contradict the claim of Exercise 9 (which says that for an i.i.d. F sequence, the expected record value sequence uniquely determines F). Note that if instead of an exchangeable sequence with common Pareto distribution (6.43) we had an i.i.d. sequence with F given by (6.43) we would have had much larger expected records. Specifically we would find

$$E(Y_n) = 2^{n+1} - 1 \tag{6.48}$$

(this follows from Exercise 8 (iii), note the present distribution differs from that discussed in the exercise by a translation of -1).

In the light of (6.47) and (6.48) it is reasonable to ask whether it is always true that for a given marginal distribution F, the i.i.d. sequence will have maximal expected records among all exchangeable sequences. This is equivalent to asking whether if

$$F(x) = \int_{-\infty}^\infty F_z(x)dF_Z(z) \tag{6.49}$$

do we have

$$F^{-1}(u) \geq \int_{-\infty}^\infty F_z^{-1}(u)dF_Z(z) \tag{6.50}$$

for large u [since the weight function $[-\log(1-u)]^n$ puts heavy weight on values of u close to 1]? Two examples are given in Exercises 11 and 12.

A second common instance of sequences of random variables with identical marginal distributions is provided by a stationary Markov chain. In this setting we have $X_1 \sim F$ and for $i > 1$,

$$F_{X_i | X_{i-1},\dots,X_1}(x_i | x_{i-1},\dots,x_1) = F_{X_i|X_{i-1}}(x_i|x_{i-1}) = \phi(x_i, x_{i-1}) \tag{6.51}$$

where, for stationarity, ϕ satisfies

$$F(x) = \int_{-\infty}^{\infty} \phi(x,y)dF(y). \tag{6.52}$$

As usual we may define a corresponding record value sequence $\{Y_n\}_{n=0}^{\infty}$. Biondini and Siddiqui (1975) point out that $\{Y_n\}$ is itself a Markov process (not stationary but with stationary transition probabilities). It is clear that

$$P(Y_n > y_n \mid Y_{n-1} = y_{n-1}) = \sum_{k=1}^{\infty} \delta_k(y_n, y_{n-1}) \tag{6.53}$$

where

$$\delta_k = P(X_j \le y_n,\ j = 2,\dots,k,\ X_{k+1} > y_n \mid X_1 = y_{n-1}).$$

Equation (6.53) can be used to verify the already known results in the i.i.d. case. For any non-trivial Markov chain it appears to be too complicated to evaluate. Biondini and Siddiqui (1975) are able to get some approximate results for a normal autoregressive scheme.

As a final example of a dependent sequence with fixed marginals we might consider a canonical maximally dependent sequence in the sense of Lai and Robbins (1978). Refer in particular to their Figure 1. If $\tilde{X}_1, \tilde{X}_2, \ldots$ is a canonical maximally dependent sequence with uniform marginals then on the probability space ([0,1], B,m) (unit interval, Borel sets and Lebesgue measure) we have

$$\tilde{Y}_0(\omega) = \tilde{X}_1(\omega) = 1 - \omega \tag{6.54}$$

while

$$\begin{aligned} \tilde{Y}_1(\omega) &= \omega, && \tfrac{1}{2} < \omega < 1 \\ &= 2\omega, && \tfrac{1}{3} < \omega < \tfrac{1}{2} \\ &= 3\omega, && \tfrac{1}{4} < \omega < \tfrac{1}{3} \\ &\text{etc.} \end{aligned} \tag{6.55}$$

Turning to the second record, again from the diagram we may determine that

$$\begin{aligned} \tilde{Y}_2(\omega) &= 2(1-\omega), && \tfrac{2}{3} < \omega < 1 \\ &= 6(\omega - \tfrac{1}{2}), && \tfrac{11}{18} < \omega < \tfrac{2}{3} \\ &= 2(1-\omega), && \tfrac{1}{2} < \omega < \tfrac{11}{18} \\ &\text{etc.} \end{aligned} \tag{6.56}$$

The expression for $\tilde{Y}_2(\omega)$ is already complicated. The prospect for writing a clear expression for $\tilde{Y}_k$ ($k > 2$) is not bright. From (6.55) we can compute the expected first record from a canonical maximally dependent uniform sequence. We find

$$E(\tilde{Y}_1) = \sum_{k=1}^{\infty} \int_{(k+1)^{-1}}^{k^{-1}} k\omega \, d\omega = \sum_{k=1}^{\infty} \frac{(2k+1)}{2k(k+1)^2}$$

and consequently

$$E(\tilde{Y}_1) = \frac{1}{2}\left[\sum_{j=1}^{\infty} j^{-2}\right] = \pi^2/12. \tag{6.57}$$

This can be compared with the corresponding result for an independent uniform sequence (from Exercise 8 (i)) namely $E(Y_1) = \frac{3}{4}$. It is clear from (6.57) that the mean first record is, as expected, larger for the maximally dependent sequence than for the i.i.d. sequence.

A canonical sequence $\{X_n\}$ of maximally dependent random variables with common distribution function F is defined by

$$X_i = F^{-1}(\tilde{X}_i) \tag{6.58}$$

where $\tilde{X}_i$ is a canonical maximally dependent uniform sequence. Using (6.55) and the fact that $Y_n = F^{-1}(\tilde{Y}_n)$, we have the following expression for the corresponding expected first record

$$E(Y_1) = \sum_{k=1}^{\infty} \int_{(k+1)^{-1}}^{k^{-1}} F^{-1}(k\omega)\, d\omega. \tag{6.59}$$

For example, if the marginal distribution is standard exponential we have $F^{-1}(u) = -\log(1-u)$. Substituting in (6.59) yields

$$\begin{aligned} E(Y_1) &= \sum_{k=1}^{\infty} \int_{(k+1)^{-1}}^{k} [-\log(1-k\omega)]d\omega \\ &= \sum_{k=1}^{\infty} \frac{1}{k} \int_0^{(k+1)^{-1}} (-\log v)dv \qquad (v = 1 - k\omega) \\ &= \sum_{k=1}^{\infty} \frac{1}{k} \int_{\log(k+1)}^{\infty} ue^{-u}du \qquad (u = -\log v) \\ &= \sum_{k=1}^{\infty} \frac{1}{k(k+1)} [1 + \log(k+1)] \\ &= 1 + \sum_{k=1}^{\infty} \frac{\log(k+1)}{k(k+1)}. \end{aligned} \tag{6.60}$$

The series displayed in (6.60) is convergent but very slowly. For example three decimal accuracy is obtainable by summing about 20,000 terms. In this fashion we have approximately for the expected first record of a canonical maximally dependent exponential sequence

$$E(Y_1) \doteq 2.257. \tag{6.61}$$

For the corresponding i.i.d. exponential sequence, the expected first record is 2. Again, as expected, the maximally dependent sequence has a larger expected first record.

For the result to hold in general we would need to verify (using (6.25) and (6.59) with convenient changes of variable) that, for any F^{-1},

$$\sum_{k=1}^{\infty} \frac{1}{k} \int_0^{(k+1)^{-1}} F^{-1}(1-u)du \geq \int_0^1 F^{-1}(1-u)(-\log u)du. \tag{6.62}$$

This inequality is studied in Exercise 14.

Exercises

1. Let $Y_1, Y_2, \dots$ be a record value sequence. Show that it constitutes a stationary Markov chain. Identify the corresponding transition probability kernel (for convenience assume the X_i's (used to define the Y_n's) are absolutely continuous with common density $f(x)$).

2. Verify (6.17) using induction and (6.15).

3. Let Y_n'' denote the n'th lower record corresponding to $X_1, X_2, \dots$ i.i.d. $\mathcal{U}(0,1)$. Verify that $Y_n'' \stackrel{d}{=} \prod_{i=0}^{n} U_i$ where the U_i's are i.i.d. $\mathcal{U}(0,1)$. Use this to derive inductively the density of Y_n''.

4. (Houchens (1984)). Let L_m and U_m denote the current values of the lower record value and upper record value sequences when the m'th record of any kind (upper or lower) is observed. Determine the joint distribution of (L_m, U_m). Hint: first consider the case where the X_i's are i.i.d. $\mathcal{U}(0,1)$. Verify inductively that in the uniform case we have for $0 < \ell < u < 1$

$$f_{L_m, U_m}(\ell, u) = 2^m[-\log(1-u+\ell)]^{m-1}/(m-1)!$$

5. Verify that if the X_i's are i.i.d. as described in (6.26) then equality obtains in (6.25).

6. Assume the X_i's have a symmetrical parent distribution with mean 0 and variance 1. Verify the bound for $E(Y_n)$ given by (6.30).

7. Let $U \sim$ uniform $(0,1)$. Define

$$X = [-\log(1-U)]^n - [-\log U]^n.$$

Determine a_n and b_n such that

$$E(a_n X + b_n) = 0, \quad E(a_n X + b_n)^2 = 1$$

and equality obtains in (6.30).

8. Suppose $X_1, X_2, \dots$ are i.i.d. F with corresponding record value sequence $\{Y_n\}_{n=0}^{\infty}$. Verify the following expressions for $E(Y_n)$:

(i) [uniform (a,b), $F(x) = \frac{x-a}{b-a}$, $a < x < b$]

$$E(Y_n) = a + (b-a)[1 - 2^{-(n+1)}]$$

(ii) [Weibull: $\bar{F}(x) = \exp[-(x/\sigma)^{\gamma}]$, $x > 0$]

$$E(Y_n) = \sigma\Gamma(n + 1 + \gamma^{-1})/\Gamma(n+1)$$

(iii) [Pareto: $\bar{F}(x) = (x/\sigma)^{-\alpha}$, $x > \sigma$, $\alpha > 1$]

$$E(Y_n) = \sigma(1 - \alpha^{-1})^{-(n+1)}.$$

9. The sequence of expected record values $E(Y_n)$, $n = 0,1,2,\ldots$ corresponding to $\{X_i\}$ i.i.d. F determines the distribution F. [Note that $E(Y_n) = \int_0^1 F^{-1}(u)\,[-\log(1-u)]^n/n!\,du$. Make the change of variable $t = -\log(1-u)$].

10. Suppose given $Z = z$, $X_1, X_2, \ldots$ are i.i.d. $F(x/z)$ (i.e. Z is a random scale parameter). Assuming appropriate expectations exist, evaluate $E(Y_n)$ the expected n'th record of the exchangeable sequence $X_1, X_2, \ldots$.

11. Suppose that given $Z = 1$, $X_1, X_2, \ldots$ are i.i.d. with common d.f. $F_1(x) = x^2$, $0 < x < 1$, while given $Z = 2$, their common distribution is $F_2(x) = 1 - (1 - x^2)$, $0 < x < 1$. Assume that $P(Z = 1) = P(Z = 2) = 1/2$. In this case $F(x) = \int_{-\infty}^{\infty} F_z(x)\,dF_Z(z) = x$, $0 < x < 1$. Verify that (6.50) holds for $u > 1/2$.

12. Suppose that given $Z = z$, $\{X_i\}$ are i.i.d. $N(z,1)$ and that $Z \sim N(0,\tau)$ for some $\tau > 0$. Verify that (6.50) holds for $u > 1/2$ in this case.

13. Determine $E(Y_1)$ the expected first record based on a canonical maximally dependent sequence with $F_{X_i}(x) = x^\gamma$, $0 < x < 1$.

14. Verify (6.62).

15. (k'th record values, Grudzien and Szynal (1983)). By keeping track of the k'th largest X yet seen, one may define a k'th record value sequence. Most of the material in this chapter can be extended to cover such sequences. The key result is that the k'th record value sequence corresponding to a distribution F is identical in distribution to the (first) record value sequence corresponding to the distribution $1 - (1 - F)^k$.

REFERENCES

ABDEL–ATY, S.H. (1954). Ordered variables in discontinuous distributions, **Statist. Neerlandica 8**, 61–82.

ABDELHAMID, S.N. (1985). On a characterization of rectangular distributions, **Statist. Prob. Letters 3**, 235–238.

ABRAMOWITZ, M. and STEGUN, I.A. (Eds.) (1965). **Handbook of Mathematical Functions with Formulas, Graphs, and Mathematical Tables**, New York: Dover Publications.

AFONJA, B. (1972). The moments of the maximum of correlated normal and t–variates, **J. Roy. Statist. Soc. B 34**, 251–262.

AITKEN, A.C. (1939). **Determinants and Matrices**, Edinburgh: Oliver & Boyd.

ALI, M.M. and CHAN, L.K. (1965). Some bounds for expected values of order statistics, **Ann. Math. Statist. 36**, 1055–1057.

ALI, M.M. and KHAN, A.H. (1987). On order statistics from the log–logistic distribution, **J. Statist. Plann. Inf. 17**, 103–108.

ANDREWS, D.F., BICKEL, P.J., HAMPEL, F.R., HUBER, P.J., ROGERS, W.H. and TUKEY, J.W. (1972). **Robust Estimates of Location**, Princeton University Press.

ARNOLD, B.C. (1974). Schwarz, regression and extreme deviance, **Amer. Statist. 28**, 22–23.

ARNOLD, B.C. (1977). Recurrence relations between expectations of functions of order statistics, **Scand. Actuar. J. 1977**, 169–174.

ARNOLD, B.C. (1980). Distribution–free bounds on the mean of the maximum of a dependent sample, **SIAM J. Appl. Math.** 38, 163–167.

ARNOLD, B.C. (1985). p–Norm bounds on the expectation of the maximum of possibly dependent sample, **J. Multivar. Anal. 17**, 316–332.

ARNOLD, B.C. and BROCKETT, P. (1988). Variance bounds using a theorem of Polya, **Statist. Prob. Letters 6**, 321–326.

ARNOLD, B.C. and GROENEVELD, R.A. (1974). Bounds for deviations between sample population statistics, **Biometrika** 61, 387–389.

ARNOLD, B.C. and GROENEVELD, R.A. (1978). Bounds on deviations of estimates arising in finite population regression models, **Commun. Statist. A7**, 1173–1179.

ARNOLD, B.C. and GROENEVELD, R.A. (1979). Bounds on expectations of linear systematic statistics based on dependent samples, **Ann. Statist.** 7, 220–223. Correction **8**, 1401.

ARNOLD, B.C. and GROENEVELD, R.A. (1981). Maximal deviation between sample and population means in finite populations, **J. Amer. Statist. Assoc.** 76, 443–445.

ARNOLD, B.C. and MEEDEN, G. (1975). Characterization of distributions by sets of moments of order statistics, **Ann. Statist.** 3, 754–758.

AVEN, T. (1985). Upper (lower) bounds on the mean of the maximum (minimum) of a number of random variables, J. Appl. **Probab.** 22, 723–728.

BALAKRISHNAN, N. (1980). **Moment Problems in Order Statistics**, Ph.D. Thesis, Indian Institute of Technology, Kanpur, India.

BALAKRISHNAN, N. (1982). A note on sum of the sub–diagonal product moments of order statistics, **J. Statist. Res. 16**, 37–42.

BALAKRISHNAN, N. (1984). Algorithm AS 200. Approximating the sum of squares of normal scores, **J. Roy. Statist. Soc., Series C 33**, 242–245.

BALAKRISHNAN, N. (1985). Order statistics from the half logistic distribution, **J. Statist. Comput. Simul. 20**, 287–309.

BALAKRISHNAN, N. (1986). Order statistics from discrete distributions, **Commun. Statist. – Theor. Meth., 15(3)**, 657–675.

BALAKRISHNAN, N. (1987a). A note on moments of order statistics from exchangeable variates, **Commun. Statist. – Theor. Meth., 16(3)**, 855–861.

BALAKRISHNAN, N. (1987b). Two identities involving order statistics in the presence of an outlier, **Commun. Statist. – Theor. Meth., 16(8)**, 2385–2389.

BALAKRISHNAN, N. (1988a). Relations and identities for the moments of order statistics from a sample containing a single outlier, **Commun. Statist. – Theor. Meth., 17(7)**, 2173–2190.

BALAKRISHNAN, N. (1988b). Recurrence relations among moments of order statistics from two related outlier models. To appear in **Biometrical Journal.**

BALAKRISHNAN, N. (1988c). Recurrence relations for order statistics from n independent and non–identically distributed random variables. **Ann. Inst. Statist. Math. 40**, 273–277.

BALAKRISHNAN, N. (1989a). A relation for the covariances of order statistics from n independent and non–identically distributed random variables. To appear in **Statist. Hefte.**

BALAKRISHNAN, N. (1989b). Recurrence relations among moments of order statistics from two related sets of independent and non–identically distributed random variables. To appear in **Ann. Inst. Statist. Math.**

BALAKRISHNAN, N. (1989c). On a generalized triangle rule for order statistics and some extensions. **Submitted for publication.**

BALAKRISHNAN, N. and AMBAGASPITIYA, R.S. (1988). Relationships among moments of order statistics in samples from two related outlier models and some applications, **Commun. Statist. – Theor. Meth., 17(7)**, 2327–2341.

BALAKRISHNAN, N. and JOSHI, P.C. (1983). Single and product moments of order statistics from symmetrically truncated logistic distribution, **Demonstratio Mathematica 16**, 833–841.

BALAKRISHNAN, N. and JOSHI, P.C. (1984). Product moments of order statistics from doubly truncated exponential distribution, **Naval Res. Logist. Quart.** 31, 27–31.

BALAKRISHNAN, N. and JOSHI, P.C. (1985). Bounds for the mean of second largest order statistic in large samples, **Math. Operationsforsch. und Statist., Ser. Statist. 16**, 457–464.

BALAKRISHNAN, N. and KOCHERLAKOTA, S. (1985). On the double Weibull distribution: Order statistics and estimation, **Sankhya B** 47, 161–178.

BALAKRISHNAN, N. and KOCHERLAKOTA, S. (1986). On the moments of order statistics from doubly truncated logistic distribution, **J. Statist. Plann. Inf. 13**, 117–129.

BALAKRISHNAN, N. and MALIK, H.J. (1985). Some general identities involving order statistics, **Commun. Statist. – Theor. Meth., 14(2)**, 333–339.

BALAKRISHNAN, N. and MALIK, H.J. (1986a). A note on moments of order statistics, **Amer. Statist. 40**, 147–148.

BALAKRISHNAN, N. and MALIK, H.J. (1986b). Order statistics from the linear–exponential distribution, Part I: Incresing hazard rate case, **Commun. Statist. – Theor. Meth., 15(1)**, 179–203.

BALAKRISHNAN, N. and MALIK, H.J. (1987). Moments of order statistics from truncated log–logistic distribution, **J. Statist. Plann. Inf. 17**, 251–267.

BALAKRISHNAN, N., MALIK, H.J. and AHMED, S.E. (1988). Recurrence relations and identities for moments of order statistics, II: Specific continuous distributions, **Commun. Statist. – Theor. Meth.**

(Statist. Reviews), 17(8), 2657–2694.

BALAKRISHNAN, N., MALIK, H.J. and PUTHENPURA, S. (1987). Best linear unbiased estimation of location and scale parameters of the log–logistic distribution, **Commun. Statist. – Theor. Meth.**, **16**(12), 3477–3495.

BALAKRISHNAN, N. and PUTHENPURA, S. (1986). Best linear unbiased estimators of location and scale parameters of the half logistic distribution, **J. Statist. Comput. Simul. 25**, 193–204.

BARLOW, R.E. and PROSCHAN, F. (1966). Inequalities for linear combinations of order statistics from restricted families, **Ann. Math. Statist. 37**, 1574–1591.

BARNETT, V.D. (1966). Order statistics estimators of the location of the Cauchy distribution, **J. Amer. Statist. Assoc. 61**, 1205–1218. Correction **63**, 383–385.

BARNETT, V. and LEWIS, T. (1978). **Outliers in Statistical Data**, John Wiley & Sons, New York.

BASU, D. (1955). On statistics independent of a complete sufficient statistic, **Sankhya 15**, 377–380.

BEESACK, P.R. (1973). On bounds for the range of ordered variates, **Publications of the Electro–technical Faculty of Belgrade Univ., Math. & Phys. Series, 428**, 93–96.

BEESACK, P.R. (1976). On bounds for the range of ordered variates II, **Aequationes Math.** 14, 293–301.

BHATTACHARYYA, B.B. (1970). Reverse submartingale and some functions of order statistics, **Ann. Math. Statist.** 41, 2155–2157.

BIONDINI, R. and SIDDIQUI, M.M. (1975). Record values in Markov sequences, In: **Statistical Inference and Related Topics**, Vol. 2 (Ed., M.L. Puri), 291–352, Academic Press, New York.

BLAND, R.P. and OWEN, D.B. (1966). A note on singular normal distributions, **Ann. Inst. Statist. Math. 18**, 113–116.

BLOM, G. (1958). **Statistical Estimates and Transformed Beta–Variables**, Almqvist and Wiksell, Uppsala, Sweden.

BORENIUS, G. (1966). On the limit distribution of an extreme value in a sample from a normal distribution, **Skand. Aktuarietidskr.**, 1–15.

BOYD, A.V. (1971). Bounds for order statistics, **Publications of the Electrotechnical Faculty of Belgrade Univ., Math. & Phys. Series, 365**, 31–32.

BRUNK, H.D. (1959). Note on two papers of K.R. Nair, **J. of Indian Soc. of Agricult. Statist. 11**, 186–189.

BURROWS, P.M. (1972). Expected selection differentials for directional selection, **Biometrics 28**, 1091–1100.

BURROWS, P.M. (1975). Variance of selection differentials in normal samples, **Biometrics 31**, 125–133.

CADWELL, J.H. (1953). The distribution of quasi–ranges in samples from a normal population, **Ann. Math. Statist.** 24, 603–613.

CARLSON, P.G. (1958). A recurrence formula for the mean range for odd sample sizes, **Skand. Aktuarietidskr. 41**, 55–56.

CHANDLER, K.N. (1952). The distribution and frequency of record values, **J. Roy. Statist. Soc., Ser. B** 14, 220–228.

CHU, J.T. and HOTELLING, H. (1955). The moments of the sample median, Ann. **Math. Statist. 26**, 593–606.

CLARK, C.E. and WILLIAMS, G.T. (1958). Distributions of the members of an ordered sample, **Ann. Math. Statist.** 29, 862–870.

COHN, R., MOSTELLER, F., PRATT, J.W. and TATSUOKA, A. (1960). Maximizing the probability that adjacent order statistics of samples from several populations form overlapping intervals, Ann. Math. Statist. 31, 1095–1104.

COLE, R.H. (1951). Relations between moments of order statistics, **Ann. Math. Statist. 22**, 308–310.

CONOVER, W.J. (1965). A k–sample model in order statistics, **Ann. Math. Statist. 36**, 1223–1235.

COX, D.R. (1954). The mean and coefficient of variation of range in small samples from non–normal populations, **Biometrika 41**, 469–481. Correction **42**, 277.

CROW, E.L. and SIDDIQUI, M.M. (1967). Robust estimation of location, **J. Amer. Statist. Assoc.** 62, 353–389.

DAVID, F.N. and JOHNSON, N.L. (1954). Statistical treatment of censored data. I. Fundamental formulae, **Biometrika 41**, 228–240.

DAVID, H.A. (1966). A note on "A k–sample model in order statistics" by W.J. Conver, Ann. Math. Statist. 37, 287–288.

DAVID, H.A. (1973). Waiting time paradoxes and order statistics, **J. Amer. Statist. Assoc. 68**, 743–745.

DAVID, H.A. (1981). **Order Statistics**, Second edition, John Wiley & Sons, New York.

DAVID, H.A. (1986). Inequalities for ordered sums, **Ann. Inst. Statist. Math. 38**, 551–555.

DAVID, H.A., HARTLEY, H.O. and PEARSON, E.S. (1954). The distribution of the ratio, in a single normal sample, of range to standard deviation, **Biometrika** 41, 482–493.

DAVID, H.A. and JOSHI, P.C. (1968). Recurrence relations between moments of order statistics for exchangeable variates, **Ann. Math. Statist.** 39, 272–274.

DAVID, H.A., KENNEDY, W.J. and KNIGHT, R.D. (1977). Means, variances, and covariances of normal order statistics in the presence of an outlier, **Selected Tables in Mathematical Statistics** 5, 75–204.

DAVID, H.A. and MISHRIKY, R.S. (1968). Order statistics for discrete populations and for grouped samples, **J. Amer. Statist. Assoc.** 63, 1390–1398.

DAVID, H.A. and SHU, V.S. (1978). Robustness of location estimators in the presence of an outlier. In: DAVID, H.A. (Ed.), **Contributions to Survey Sampling and Applied Statistics: Papers in Honour** of **H.O. Hartley**, 235–250, Academic Press, New York.

DAVIS, C.S. and STEPHENS, M.A. (1977). The covariance matrix of normal order statistics, **Commun. Statist. B** 6, 75–81.

DAVIS, C.S. and STEPHENS, M.A. (1978). Approximating the covariance matrix of normal order statistics, Algorithm AS 128, **Appl. Statist.** 27, 206–212.

DAVIS, H.T. (1935). **Tables of the Higher Mathematical Functions**, Vols. 1 and 2, Bloomington: Principia Press.

DE FINETTI, B. (1937). La prevision: ses lois logiques, ses sources subjectives, **Ann. Inst. H. Poincare (Paris)** 7, 1–68.

DOWNTON, F. (1966). Linear estimates with polynomial coefficients, **Biometrika** 53, 129–141.

DWASS, M. (1975). The extreme deviations inequality (Letter), **Amer. Statist.** 29, 108.

DYER, D.D. and WHISENAND, C.W. (1973a). Best linear unbiased estimator of the parameter of the Rayleigh distribution – I: Small sample theory for censored order statistics, **IEEE Trans. Reliab. R–22**, 27–34.

DYER, D.D. and WHISENAND, C.W. (1973b). Best linear unbiased estimator of the parameter of the

Rayleigh distribution – II: Optimum theory for selected order statistics, **IEEE Trans. Reliab. R–22**, 229–231.

DYKSTRA, R.L., HEWETT, J.E. and THOMPSON, W.A., Jr. (1973). Events which are almost dependent, **Ann. Statist. 1**, 674–681.

FAHMY, S. and PROSCHAN, F. (1981). Bounds on differences of order statistics, **Amer. Statist. 35**, 46–47.

GALAMBOS, J. (1974/75). Methods for proving Bonferroni type inequalities, **J. London Math. Soc. 9**, 561–564.

GALAMBOS, J. (1978). **The Asymptotic Theory of Extreme Order Statistics**, John Wiley & Sons, New York.

GALAMBOS, J. (1979). Review of Lai and Robbins paper, **Math. Reviews 57**, #17778.

GALLOT, S. (1966). A bound for the maximum of number of random variables, **J. Appl. Prob. 3**, 556–558.

GALTON, F. (1902). The most suitable proportion between the values of first and second prizes, **Biometrika 1**, 385–390.

GASTWIRTH, J.L. and COHEN, M.L. (1970). Small sample behaviour of robust linear estimators of location, **J. Amer. Statist. Assoc. 65**, 946–973.

GILSTEIN, C.Z. (1981). Bounds for expectations of linear combinations of order statistics (Preliminary Report) (Abstract) (177–100), **Inst. Math. Statist. Bull. 10**, 253.

GODWIN, H.J. (1949). Some low moments of order statistics, **Ann. Math. Statist. 20**, 279–285.

GOEBEL, J.J. (1974). Personal communication.

GOVINDARAJULU, Z. (1963a). On moments of order statistics and quasi–ranges from normal populations, **Ann. Math. Statist.** 34, 633–651.

GOVINDARAJULU, Z. (1963b). Relationships among moments of order statistics in samples from two related populations, **Technometrics 5**, 514–518.

GOVINDARAJULU, Z. (1966). Best linear estimates under symmetric censoring of the parameters of a double exponential population, **J. Amer. Statist. Assoc.** 61, 248–258.

GOVINDARAJULU, Z. (1968). Certain general properties of unbiased estimates of location and scale parameters based on ordered observations, **SIAM J. Appl. Math. 16**, 533–551.

GRAVEL, R. and VAN EEDEN, C. (1981). Best linear unbiased estimators of the location of a double quadratic distribution based on order statistics, **J. Statist. Comput. Simul.** 13, 225–243.

GRAVEY, A. (1985). A simple construction of an upper bound for the mean of the maximum of n identically distributed random variables, **J. Appl. Prob. 22**, 844–851.

GREIG, M. (1967). Extremes in a random assembly, **Biometrika 54**, 273–282.

GROENEVELD, R.A. (1982). Best bounds for order statistics and their expectations in range and mean units with applications, **Commun. Statist. – Theor. Meth.**, 11, 1809–1815.

GRUDZIEN, A. and SZYNAL, D. (1983). On the expected values of K–th record values and associated characterizations of distributions, **Proc. of 4th Pannonian Symp. on Math. Statist., Bad Tatzmanndorf, Austria**, F. Konecny et al., eds., 119–127.

GUMBEL, E.J. (1954). The maxima of the mean largest value and of the range, **Ann. Math. Statist. 25**, 76–84.

GUPTA, S.S. (1960). Order statistics from the gamma distribution, **Technometrics 2**, 243–262.

GUPTA, S.S. (1963). Probability integrals of multivariate normal and multivariate t, **Ann. Math. Statist. 34**, 792–828.

GUTERMAN, H.E. (1962). An upper bound for the sample standard deviation, **Technometrics** 4, 134–135.

HAMPEL, F.R. (1974). The influence curve and its role in robust estimation, **J. Amer. Statist. Assoc.** 69, 383–393.

HARTER, H.L. (1961). Expected values of normal order statistics, **Biometrika 48**, 151–165. Correction **48**, 476.

HARTER, H.L. (1970a,b). **Order Statistics and their Uses in Testing and Estimation**, Vols. 1 and 2, U.S. Govt. Printing Office, Washington, D.C.

HARTLEY, H.O. and DAVID, H.A. (1954). Universal bounds for mean range and extreme observation, **Ann. Math. Statist.** 25, 85–99.

HAWKINS, D.M. (1971). On the bounds of the range of order statistics, **J. Amer. Statist. Assoc.** 66, 644–645.

HAWKINS, D.M. (1979). Fractiles of an extended multiple outlier test, **J. Statist. Comput. Simul. 8**, 227–236.

HEATH, D. and SUDDERTH, W. (1976). De Finetti's theorem on exchangeable variables, **Amer. Statist. 30**, 188–189.

HILL, W.G. (1976). Order statistics of correlated variables and implications in genetic selection programmes, **Biometrics** 32, 889–902.

HIRAKAWA, K. (1973). Moments of order statistics, **Ann. Statist. 1**, 392–394.

HOEFFDING, W. (1953). On the distribution of the expected values of the order statistics, **Ann. Math. Statist. 24**, 93–100.

HOEFFDING, W. (1956). On the distribution of the number of successes in independent trials, **Ann. Math. Statist. 27**, 713–721.

HOFFMAN, T.R. and SAW, J.G. (1975). Distribution of the largest of a set of equicorrelated normal variables, **Commun. Statist. 4**, 49–55.

HOTELLING, H. and SOLOMONS, L.M. (1932). The limits of a measure of skewness, **Ann. Math. Statist.** 3, 141–142.

HOUCHENS, R.L. (1984). **Record Value Theory and Inference**, Unpublished Ph.D. Dissertation, University of California, Riverside, California.

JONES, H.L. (1948). Exact lower moments of order statistics in small samples from a normal distribution, **Ann. Math. Statist.** 19, 270–273.

JOSHI, P.C. (1969). Bounds and approximations for the moments of order statistics, **J. Amer. Statist. Assoc.** 64, 1617–1624.

JOSHI, P.C. (1971). Recurrence relations for the mixed moments of order statistics, **Ann. Math. Statist.** 42, 1096–1098.

JOSHI, P.C. (1972). Efficient estimation of the mean of an exponential distribution when an outlier is present, **Technometrics** 14, 137–144.

JOSHI, P.C. (1973). Two identities involving order statistics, **Biometrika 60**, 428–429.

JOSHI, P.C. (1978). Recurrence relations between moments of order statistics from exponential and truncated exponential distributions, **Sankhya B 39**, 362–371.

JOSHI, P.C. (1979a). A note on the moments of order statistics from doubly truncated exponential distribution, **Ann. Inst. Statist. Math.** 31, 321–324.

JOSHI, P.C. (1979b). On the moments of gamma order statistics, **Naval Res. Logist. Quart.** 26, 675–679.

JOSHI, P.C. (1982). A note on the mixed moments of order statistics from exponential and truncated exponential distributions, **J. Statist. Plann. Inf.** 6, 13–16.

JOSHI, P.C. and BALAKRISHNAN, N. (1981a). Applications of order statistics in combinatorial identities, **J. Comb. Infor. System Sci.** 6, 271–278.

JOSHI, P.C. and BALAKRISHNAN, N. (1981b). An identity for the moments of normal order statistics with applications, **Scand. Actuarial J.** 1981, 203–213.

JOSHI, P.C. and BALAKRISHNAN, N. (1982). Recurrence relations and identities for the product moments of order statistics, **Sankhya B** 44, 39–49.

JOSHI, P.C. and BALAKRISHNAN, N. (1983). Bounds for the moments of extreme order statistics for large samples, **Math. Operationsforsch. und Statist., Ser. Statist.** 14, 387–396.

KABE, D.G. (1980). On extensions of Samuelson's inequality (Letter), **Amer. Statist.** 34, 249.

KABIR, A.B.M.L. and RAHMAN, M. (1974). Bounds for expected values of order statistics, **Commun. Statist.** 3, 557–566.

KAIGH, W.D. (1980). An empirical regression inequality, **Biomet. J.** 22, 83–85.

KALE, B.K. and SINHA, S.K. (1971). Estimation of expected life in the presence of an outlier observation, **Technometrics** 13, 755–759.

KARLIN, S. (1968). **Total Positivity**, Stanford University Press, Stanford, California.

KELLEHER, G.J. and WALSH, J.E. (1972). Exact intervals and tests for median when one "sample" value possibly an outlier, **Trabajos de Estadistica** 23, 121–124.

KEMPTHORNE, O. (1973). Personal Communication.

KHATRI, C.G. (1962). Distributions of order statistics for discrete case, **Ann. Inst. Statist. Math.** 14, 167–171.

KIMBALL, B.F. (1947). Assignment of frequencies to a completely ordered set of sample data (discussion), **Trans. Amer. Geophysicists Union** 28, 952.

KLAMKIN, M.S. (1974). An inequality of statistical interest, Problem E2428, **Amer. Math. Monthly** 81, 782–783.

KOOP, J.C. (1972). On the derivation of expected value and variance of ratios without the use of infinite series expansions, **Metrika** 19, 156–170.

KRISHNAIAH, P.R. and RIZVI, M.H. (1966). A note on recurrence relations between expected values of functions of order statistics, **Ann. Math. Statist.** 37, 733–734.

LAI, T.L. and ROBBINS, H. (1976). Maximally dependent random variables, **Proc. Nat. Acad. Sci. U.S.A.** 73, 286–288.

LAI, T.L. and ROBBINS, H. (1978). A class of dependent random variables and their maxima, **Z. Wahrsch. Verw. Gebiete** 42, 89–111.

LAWLESS, J.F. (1982). **Statistical Models & Methods for Lifetime Data**, John Wiley & Sons, New York.

LEFEVRE, C. (1986). Bounds for the expectation of linear combinations of order statistics with application to Pert networks, **Stoch. Anal. Appl.** 4, 351–356.

LIEBLEIN, J. (1955). On moments of order statistics from the Weibull distribution, **Ann. Math. Statist.** 26, 330–333.

LOEVE, M. (1977). **Probability Theory** I, 4th edition, Springer–Verlag, New York.

LOYNES, R.M. (1979). A note on Prescott's upper bounds for normed residuals, **Biometrika 66**, 387–389.

MAJINDAR, K.N. (1962). Improved bounds on a measure of skewness, **Ann. Math. Statist. 33**, 1192–1194.

MALIK, H.J. (1967). Exact moments of order statistics from a power–function distribution, **Skand. Aktuarietidskr. 1967**, 64–69.

MALIK, H.J., BALAKRISHNAN, N. and AHMED, S.E. (1988). Recurrence relations and identities for moments of order statistics, I: Arbitrary continuous distribution, **Commun. Statist. – Theor.** Meth. **(Statist. Reviews), 17(8)**, 2623–2655.

MALLOWS, C.L. (1969). Extrema of expectations of uniform order statistics, Problem 68–5, **SIAM Review 10, 109**. Solution 11, 410–411.

MALLOWS, C.L. and RICHTER, D. (1964). Sharp bounds for two measures of skewness, **Ann. Math. Statist. 35**, 460–461.

MALLOWS, C.L. and RICHTER, D. (1969). Inequalities of Chebyshev type involving conditional expectations, **Ann. Math. Statist. 40**, 1922–1932.

MARGOLIN, B.H. and WINOKUR, H.S., Jr. (1967). Exact moments of the order statistics of the geometric distribution and their relation to inverse sampling and reliability of reduntant systems, **J. Amer. Statist. Assoc. 62**, 915–925.

MATHAI, A.M. (1975). Bounds for moments through a general othogonal expansion in a pre–Hilbert space I, **Canad. J. Statist. 3**, 13–34.

MATHAI, A.M. (1976). Bounds for moments through a general orthogonal expansion in a pre–Hilbert space II, **Canad. J. Statist. 4**, 1–12.

MAURER, W. and MARGOLIN, B.H. (1976). The multivariate inclusion–exclusion formula and order statistics from dependent variates, **Ann. Statist. 4**, 1190–1199.

McKAY, A.T. (1935). The distribution of the difference between the extreme observation and the sample mean of n from a normal universe, **Biometrika 27**, 466–471.

MELNICK, E.L. (1964). **Moments of Ranked Poisson Variates**, M.S. Thesis, Virginia Polytechnic Institute.

MENDENHALL, W. (1983). **Introduction to Probability and Statistics**, 6th edition, Duxbury Press, Boston.

MILTON, R.C. (1970). **Rank Order Probability: Two–Sample Normal Shift Alternatives**, John Wiley & Sons, New York.

MORIGUTI, S. (1951). Extremal properties of extreme value distributions, **Ann. Math. Statist. 22**, 523–536.

MORIGUTI, S. (1953). A modification of Schwarz's inequality with applications to distributions, **Ann. Math. Statist. 24**, 107–113.

MURPHY, R.B. (1951). **On Tests For Outlying Observations**, Ph.D. Thesis, Princeton University.

MUSTAFI, C.K. (1969). A recurrence relation for distribution functions of order statistics from bivariate distributions, **J. Amer. Statist. Assoc. 64**, 600–601.

NAGARAJA, H.N. (1978). On the expected values of record values, **Austral. J. Statist. 20**, 176–182.

NAGARAJA, H.N. (1979). Personal Communication.

NAGARAJA, H.N. (1981). Some finite sample results for the selection differential, **Ann. Inst. Statist. Math. 33**, 437–448.

NAIR, K.R. (1948). The distribution of the extreme deviate from the sample mean and its studentized form, **Biometrika** 35, 118–144.

NEYMAN, J. and SCOTT, E.L. (1971). Outlier proneness of phenomena and of related distributions. In: Rustagi, J. (Ed.), **Optimizing Methods in Statistics**, 413–430. Academic Press, New York.

O'REILLY, F.J. (1975). On a criterion for extrapolation in normal regression, **Ann. Statist.** 3, 219–222.

O'REILLY, F.J. (1976). The extreme deviations inequality (Letter), **Amer. Statist.** 30, 103.

PATEL, J.K. (1975). Bounds on moments of linear functions of order statistics from Weibull and other restricted families, **J. Amer. Statist. Assoc. 70**, 670–672.

PETEL, J.K. and READ, C.B. (1975). Bounds on conditional moments of Weibull and other monotone failure rate families, **J. Amer. Statist. Assoc. 70**, 238–244.

PEARSON, E.S. and CHANDRA SEKAR, C. (1936). The efficiency of statistical tools and a criterion for the rejection of outlying observations, **Biometrika 28**, 308–319.

PEARSON, K. (1902). Note on Francis Galton's difference problem, **Biometrika** 1, 390–399.

PEARSON, K. and PEARSON, M.V. (1931). On the mean character and variance of a ranked individual, and on the mean and variance of the intervals between ranked individuals. I (1931): Symmetrical distributions (normal and rectangular), **Biometrika** 23, 364–397. II (1932): Case of certain skew curves, **Biometrika 24**, 203–279.

PLACKETT, R.L. (1947). Limits of the ratio of mean range to standard deviation, **Biometrika** 34, 120–122.

PLACKETT, R.L. (1958). Linear estimation from censored data, **Ann. Math. Statist.** 29, 131–142.

PRESCOTT, P. (1977). An upper bound for any linear function of normed residuals, **Commun. Statist. B6, 83–88.**

RAGAB, A. and GREEN, J. (1984). On order statistics from the log–logistic distribution and their properties, **Commun. Statist. – Theor. Meth., 13(21)**, 2713–2724.

RAWLINGS, J.O. (1976). Order statistics for a special class of unequally correlated multinomial variates, **Biometrics 32**, 875–888.

RESNICK, S.I. (1973). Record values and maxima, **Ann. Prob.** 1, 650–662.

RIDER, P.R. (1960). Variance of the median of samples from a Cauchy distribution, **J. Amer. Statist. Assoc.** 55, 322–323.

ROHATGI, V.K. and SALEH, A.K.Md.E. (1988). A class of distributions connected to order statistics with nonintegral sample szie, **Commun. Statist. – Theor. Meth., 17(6)**, 2005–2012.

ROMANOVSKY, V. (1933). On a property of the mean ranges in samples from a normal population and on some integrals of Professor T. Hojo, **Biometrika** 25, 195–197.

ROYSTON, J.P. (1982). Algorithm AS 177. Expected normal order statistics (exact and approximate), **J. Roy. Statist. Soc., Series C** 31, 161–163.

RUBEN, H. (1956a). On the moments of the range and product moments of extreme order statistics in normal samples, **Biometrika 43**, 458–460.

RUBEN, H. (1956b). On the sum of squares of normal scores, **Biometrika 43**, 456–458. Correction 52, 669.

SALEH, A.K.Md.E., SCOTT, C. and JUNKINS, D.B. (1975). Exact first and second order moments of order statistics from the truncated exponential distribution, **Naval Res. Logist. Quart.** 22, 65–77.

SAMUELSON, P.A. (1968). How deviant can you be?, **J. Amer. Statist. Assoc.** 63, 1522–1525.

SARHAN, A.E. and GREENBERG, B.G. (1956). Estimation of location and scale parameters by order statistics from singly and doubly censored samples. Part I. The normal distribution up to samples of size 10, **Ann. Math. Statist.** 27, 427–451. Correction **40**, 325.

SAW, J.G. (1960). A note on the error after a number of terms of the David–Johnson series for the expected values of normal order statistics, **Biometrika 47**, 79–86.

SAW, J.G. and CHOW, B. (1966). The curve through the expected values of ordered variates and the sum of squares of normal scores, **Biometrika 53**, 252–255.

SCHAEFFER, L.R., VAN VLECK, L.D. and VELASCO, J.A. (1970). The use of order statistics with selected records, **Biometrics 26**, 854–859.

SCOTT, J.M.C. (1936). Appendix to paper by Pearson and Chandra Sekar, **Biometrika 28**, 319–320.

SEAL, K.C. (1956). On minimum variance among certain linear functions of order statistics, **Ann. Math. Statist. 27**, 854–855.

SEN, P.K. (1959). On moments of the sample quantiles, **Calcutta Statist. Assoc. Bull. 9**, 1–19.

SEN, P.K. (1970). A note on order statistics for heterogeneous distributions, **Ann. Math. Statist.** 41, 2137–2139.

SHAH, B.K. (1966). On the bivariate moments of order statistics from a logistic distribution, Ann. **Math. Statist. 37**, 1002–1010.

SHAH, B.K. (1970). Note on moments of a logistic order statistics, **Ann. Math. Statist. 41**, 2151–2152.

SHU, V.S. (1978). **Robust Estimation of a Location Parameter in the Presence of Outliers**, Ph.D. Thesis, Iowa State University.

SIDDIQUI, M.M. (1962). Approximations to the moments of the sample median, **Ann. Math. Statist.** 33, 157–168.

SILLITTO, G.P. (1951). Interrelations between certain linear systematic statistics of samples from any continuous population, **Biometrika** 38, 377–382.

SILLITTO, G.P. (1964). Some relations between expectations of order statistics in samples of different sizes, **Biometrika** 51, 259–262.

SIOTANI, M. (1957). Order statistics for discrete case with a numerical application to the binomial distribution, Ann. **Inst. Statist. Math. 8**, 95–104.

SMITH, N.L. and TONG, Y.L. (1983). Inequalities for functions of order statistics under an additive model, **Ann. Inst. Statist. Math. 35**, 255–265.

SMITH, W.P. (1980). Letter quoted in Editor's Note, **Amer. Statist.** 34, 251.

SRIKANTAN, K.S. (1962). Recurrence relations between the PDF's of order statistics, and some applications, **Ann. Math. Statist.** 33, 169–177.

STIGLER, S.M. (1977). Fractional order statistics, with applications, **J. Amer. Statist. Assoc. 72**, 544–550.

SUGIURA, N. (1962). On the orthogonal inverse expansion with an application to the moments of order statistics, **Osaka Math. J. 14**, 253–263.

SUGIURA, N. (1964). The bivariate orthogonal inverse expansion and the moments of order statistics, **Osaka J. Math.** 1, 45–59.

SUKHATME, P.V. (1937). Tests of significance for samples of the CHI–SQUARE population with 2 degrees of freedom, **Ann. Eugen.** 8, 52–56.

TADIKAMALLA, P.R. (1977). An approximation to the moments and the percentiles of gamma order statistics, **Sankhya B 39**, 372–381.

TEICHROEW, D. (1956). Tables of expected values of order statistics and products of order statistics for samples of size twenty and less from the normal distribution, **Ann. Math. Statist. 27**, 410–426.

THOMSON, G.W. (1955). Bounds for the ratio of range to standard deviation, **Biometrika 42**, 268–269.

TIETJEN, G.L., KAHANER, D.K. and BECKMAN, R.J. (1977). Variances and covariances of the normal order statistics for sample sizes 2 to 50, **Selected Tables in Mathematical Statistics 5**, 1–73.

TIKU, M.L. (1967). Estimating the mean and standard deviation from a censored normal sample, **Biometrika 54**, 155–165.

TIKU, M.L. (1980). Robustness of MML estimators based on censored samples and robust test statistics, **J. Statist. Plann. Inf. 4**, 123–143.

TIKU, M.L. and MALIK, H.J. (1972). On the distribution of order statistics, **Austral. J. Statist. 14**, 103–108.

TIKU, M.L., TAN, W.Y. and BALAKRISHNAN, N. (1986). **Robust Inference**, Marcel Dekker, New York.

TIPPETT, L.H.C. (1925). On the extreme individuals and the range of samples taken from a normal population, **Biometrika 17**, 364–387.

VAN ZWET, W.R. (1964). **Convex Transformations of Random Variables**, Mathematical Centre Tracts 7, Mathematisch Centrum, Amsterdam.

VAUGHAN, R.J. and VENABLES, W.N. (1972). Permanent expressions for order statistics densities, **J. Roy. Statist. Soc., Ser. B**, 34, 308–310.

WHITTLE, P. (1959). Sur la distribution du maximum d'un polynome trigonometrigue a coefficients aleatoires, **Calc. Prob. Appl. Internat. Centre Nat. Res. Sci. 87**, 173–184.

WOLKOWICZ, H. and STYAN, G.P.H. (1979). Extensions of Samuelson's inequality, **Amer. Statist. 33**, 143–144.

YAMAUTI, Z. (Ed.) (1972). **Statistical Tables and Formulas with Computer Applications**, JSA–1972, Japanese Standard Association, Tokyo, Japan.

YOUNG, D.H. (1967). Recurrence relations between the P.D.F.'s of order statistics of dependent variables, and some applications, **Biometrika 54**, 283–292.

YOUNG, D.H. (1970). The order statistics of the negative binomial distribution, **Biometrika 57**, 181–186.

YOUNG, D.H. (1971). Moment relations for order statistics of the standardized gamma distribution and the inverse multinomial distribution, **Biometrika** 58, 637–640.

AUTHOR INDEX

SUBJECT INDEX

Lecture Notes in Statistics

Vol. 1: R.A. Fisher: An Appreciation. Edited by S.E. Fienberg and D.V. Hinkley. XI, 208 pages, 1980.

Vol. 2: Mathematical Statistics and Probability Theory. Proceedings 1978. Edited by W. Klonecki, A. Kozek, and J. Rosiński. XXIV, 373 pages, 1980.

Vol. 3: B. D. Spencer, Benefit-Cost Analysis of Data Used to Allocate Funds. VIII, 296 pages, 1980.

Vol. 4: E. A. van Doorn, Stochastic Monotonicity and Queueing Applications of Birth-Death Processes. VI, 118 pages, 1981.

Vol. 5: T. Rolski, Stationary Random Processes Associated with Point Processes. VI, 139 pages, 1981.

Vol. 6: S. S. Gupta and D.-Y. Huang, Multiple Statistical Decision Theory: Recent Developments. VIII, 104 pages, 1981.

Vol. 7: M. Akahira and K. Takeuchi, Asymptotic Efficiency of Statistical Estimators. VIII, 242 pages, 1981.

Vol. 8: The First Pannonian Symposium on Mathematical Statistics. Edited by P. Révész, L. Schmetterer, and V. M. Zolotarev. VI, 308 pages, 1981.

Vol. 9: B. Jørgensen, Statistical Properties of the Generalized Inverse Gaussian Distribution. VI, 188 pages, 1981.

Vol. 10: A. A. McIntosh, Fitting Linear Models: An Application on Conjugate Gradient Algorithms. VI, 200 pages, 1982.

Vol. 11: D. F. Nicholls and B. G. Quinn, Random Coefficient Autoregressive Models: An Introduction. V, 154 pages, 1982.

Vol. 12: M. Jacobsen, Statistical Analysis of Counting Processes. VII, 226 pages, 1982.

Vol. 13: J. Pfanzagl (with the assistance of W. Wefelmeyer), Contributions to a General Asymptotic Statistical Theory. VII, 315 pages, 1982.

Vol. 14: GLIM 82: Proceedings of the International Conference on Generalised Linear Models. Edited by R. Gilchrist. V, 188 pages, 1982.

Vol. 15: K. R. W. Brewer and M. Hanif, Sampling with Unequal Probabilities. IX, 164 pages, 1983.

Vol. 16: Specifying Statistical Models: From Parametric to Non-Parametric, Using Bayesian or Non-Bayesian Approaches. Edited by J.P. Florens, M. Mouchart, J.P. Raoult, L. Simar, and A.F.M. Smith. XI, 204 pages, 1983.

Vol. 17: I.V. Basawa and D. J. Scott, Asymptotic Optimal Inference for Non-Ergodic Models. IX, 170 pages, 1983.

Vol. 18: W. Britton, Conjugate Duality and the Exponential Fourier Spectrum. V, 226 pages, 1983.

Vol. 19: L. Fernholz, von Mises Calculus For Statistical Functionals. VIII, 124 pages, 1983.

Vol. 20: Mathematical Learning Models – Theory and Algorithms: Proceedings of a Conference. Edited by U. Herkenrath, D. Kalin, W. Vogel. XIV, 226 pages, 1983.

Vol. 21: H. Tong, Threshold Models in Non-linear Time Series Analysis. X, 323 pages, 1983.

Vol. 22: S. Johansen, Functional Relations, Random Coefficients and Nonlinear Regression with Application to Kinetic Data. VIII, 126 pages. 1984.

Vol. 23: D. G. Saphire, Estimation of Victimization Prevalence Using Data from the National Crime Survey. V, 165 pages. 1984.

Vol. 24: T.S. Rao, M.M. Gabr, An Introduction to Bispectral Analysis and Bilinear Time Series Models. VIII, 280 pages, 1984.

Vol. 25: Time Series Analysis of Irregularly Observed Data. Proceedings, 1983. Edited by E. Parzen. VII, 363 pages, 1984.

ctd. on inside back cover

Lecture Notes in Statistics

Vol. 26: Robust and Nonlinear Time Series Analysis. Proceedings, 1983. Edited by J. Franke, W. Härdle and D. Martin. IX, 286 pages. 1984.

Vol. 27: A. Janssen, H. Milbrodt, H. Strasser, Infinitely Divisible Statistical Experiments. VI, 163 pages. 1985.

Vol. 28: S. Amari, Differential-Geometrical Methods in Statistics. V, 290 pages. 1985.

Vol. 29: Statistics in Ornithology. Edited by B.J.T. Morgan and P.M. North. XXV, 418 pages. 1985.

Vol. 30: J. Grandell, Stochastic Models of Air Pollutant Concentration. V, 110 pages. 1985.

Vol. 31: J. Pfanzagl, Asymptotic Expansions for General Statistical Models. VII, 505 pages. 1985.

Vol. 32: Generalized Linear Models. Proceedings, 1985. Edited by R. Gilchrist, B. Francis and J. Whittaker. VI, 178 pages. 1985.

Vol. 33: M. Csörgő, S. Csörgő, L. Horváth, An Asymptotic Theory for Empirical Reliability and Concentration Processes. V, 171 pages. 1986.

Vol. 34: D.E. Critchlow, Metric Methods for Analyzing Partially Ranked Data. X, 216 pages. 1985.

Vol. 35: Linear Statistical Inference. Proceedings, 1984. Edited by T. Caliński and W. Klonecki. VI, 318 pages. 1985.

Vol. 36: B. Matérn, Spatial Variation. Second Edition. 151 pages. 1986.

Vol. 37: Advances in Order Restricted Statistical Inference. Proceedings, 1985. Edited by R. Dykstra, T. Robertson and F.T. Wright. VIII, 295 pages. 1986.

Vol. 38: Survey Research Designs: Towards a Better Understanding of Their Costs and Benefits. Edited by R.W. Pearson and R.F. Boruch. V, 129 pages. 1986.

Vol. 39: J.D. Malley, Optimal Unbiased Estimation of Variance Components. IX, 146 pages. 1986.

Vol. 40: H.R. Lerche, Boundary Crossing of Brownian Motion. V, 142 pages. 1986.

Vol. 41: F. Baccelli, P. Brémaud, Palm Probabilities and Stationary Queues. VII, 106 pages. 1987.

Vol. 42: S. Kullback, J.C. Keegel, J.H. Kullback, Topics in Statistical Information Theory. IX, 158 pages. 1987.

Vol. 43: B.C. Arnold, Majorization and the Lorenz Order: A Brief Introduction. VI, 122 pages. 1987.

Vol. 44: D.L. McLeish, Christopher G. Small, The Theory and Applications of Statistical Inference Functions. 136 pages. 1987.

Vol. 45: J.K. Ghosh, Statistical Information and Likelihood. 384 pages. 1988.

Vol. 46: H.-G. Müller, Nonparametric Regression Analysis of Longitudinal Data. VI, 199 pages. 1988.

Vol. 47: A.J. Getson, F.C. Hsuan, {2}-Inverses and Their Statistical Application. VIII, 110 pages. 1988.

Vol. 48: G.L. Bretthorst, Bayesian Spectrum Analysis and Parameter Estimation. XII, 209 pages. 1988.

Vol. 49: S.L. Lauritzen, Extremal Families and Systems of Sufficient Statistics. XV, 268 pages. 1988.

Vol. 50: O.E. Barndorff-Nielsen, Parametric Statistical Models and Likelihood. VII, 276 pages. 1988.

Vol. 51: J. Hüsler, R.-D. Reiss (Eds.), Extreme Value Theory. Proceedings, 1987. X, 279 pages. 1989.

Vol. 52: P.K. Goel, T. Ramalingam, The Matching Methodology: Some Statistical Properties. VIII, 152 pages. 1989.

Vol. 53: B.C. Arnold, N. Balakrishnan, Relations, Bounds and Approximations for Order Statistics. IX, 173 pages. 1989.